URBAN BIODIVERSITY

Urban biodiversity is an increasingly popular topic among researchers. Worldwide, thousands of research projects are unravelling how urbanisation impacts the biodiversity of cities and towns, as well as its benefits for people and the environment through ecosystem services. Exciting scientific discoveries are made on a daily basis. However, researchers often lack time and opportunity to communicate these findings to the community and those in charge of managing, planning and designing for urban biodiversity. On the other hand, urban practitioners frequently ask researchers for more comprehensible information and actionable tools to guide their actions.

This book is designed to fill this cultural and communicative gap by discussing a selection of topics related to urban biodiversity, as well as its benefits for people and the urban environment. It provides an interdisciplinary overview of scientifically grounded knowledge vital for current and future practitioners in charge of urban biodiversity management, its conservation and integration into urban planning. Topics covered include pests and invasive species, rewilding habitats, the contribution of a diverse urban agriculture to food production, implications for human well-being, and how to engage the public with urban conservation strategies.

For the first time, world-leading researchers from five continents convene to offer a global interdisciplinary perspective on urban biodiversity narrated with a simple but rigorous language. This book synthesizes research at a level suitable for both students and professionals working in nature conservation and urban planning and management.

Alessandro Ossola is a Research Coordinator within the Centre for Smart Green Cities at Macquarie University, Sydney, Australia. He is also an Honorary Research Fellow at the University of Melbourne, Australia and a former US National Academy of Science, Engineering and Medicine NRC Associate within the US Environmental Protection Agency in Cincinnati, Ohio, USA.

Jari Niemelä is Professor of Urban Ecology and Director of the Helsinki Institute for Sustainability Science, Department of Environmental Sciences, at the University of Helsinki, Finland.

Routledge Studies in Urban Ecology

URBAN BIODIVERSITY

From Research to Practice

Edited by Alessandro Ossola and Jari Niemelä

Routledge
Taylor & Francis Group

earthscan
from Routledge

LONDON AND NEW YORK

First published 2018
by Routledge
2 Park Square, Milton Park, Abingdon, Oxon OX14 4RN

and by Routledge
711 Third Avenue, New York, NY 10017

Routledge is an imprint of the Taylor & Francis Group, an informa business

British Library Cataloguing-in-Publication Data
A catalogue record for this book is available from the British Library

Library of Congress Cataloging-in-Publication Data
Names: Ossola, Alessandro, editor. | Niemelèa, Jari, editor.
Title: Urban biodiversity : from research to practice / edited by
    Alessandro Ossola and Jari Niemelèa.
Description: Milton Park, Abingdon, Oxon ; New York, NY : Routledge,
    2018. | Series: Routledge studies in urban ecology | Includes
    bibliographical references and index.
Identifiers: LCCN 2017029223| ISBN 9781138224384 (hbk) | ISBN
    9781138224391 (pbk) | ISBN 9781315402581 (ebook)
Subjects: LCSH: Urban ecology (Biology)—Research.
Classification: LCC QH541.5.C6 U73 2018 | DDC 307.76072—dc23
LC record available at https://lccn.loc.gov/2017029223

ISBN: 978-1-138-22438-4 (hbk)
ISBN: 978-1-138-22439-1 (pbk)
ISBN: 978-1-315-40258-1 (ebk)

Typeset in Bembo
by Swales & Willis Ltd, Exeter, Devon, UK

CONTENTS

CONTRIBUTORS

Elsa C. Anderson, University of Illinois, Chicago, IL, USA.

Myla F.J. Aronson, Department of Ecology, Evolution and Natural Resources, Rutgers – The State University of New Jersey, New Brunswick, NJ, USA.

Francisco de la Barrera, Institute of Geography, Center for Sustainable Urban Development (CEDEUS), Pontificia Universidad Catolica de Chile, Santiago, Chile.

J. Amy Belaire, St. Edward's University, Austin, TX, USA.

Sara Borgström, Royal Institute of Technology, KTH, Department of Sustainable Development, Environmental Science and Engineering, Stockholm, Sweden.

Sarel S. Cilliers, Unit for Environmental Sciences and Management, North-West University, Potchefstroom, South Africa.

Elandrie Davoren, Unit for Environmental Sciences and Management, North-West University, Potchefstroom, South Africa.

Cynnamon Dobbs, Department of Ecosystems and Environment, Faculty of Agronomy and Forest Engineering, Pontificia Universidad Catolica de Chile, Santiago, Chile.

Marié J. Du Toit, Unit for Environmental Sciences and Management, North-West University, Potchefstroom, South Africa.

Monika H. Egerer, Environmental Studies Department, University of California, Santa Cruz, CA, USA.

Megan Garfinkel, University of Illinois, Chicago, IL, USA.

Erik Gómez-Baggethun, Department of International Environment and Development Studies (Noragric), Norwegian University of Life Sciences (NMBU), Ås, Norway, and Norwegian Institute for Nature Research (NINA).

Angela Hernández, Centro de Investigacion en Ecosistemas de la Patagonia (CIEP), Coyhaique, Chile.

Dieter F. Hochuli, School of Life and Environmental Sciences, The University of Sydney, NSW, Australia.

Maria Ignatieva, Swedish University of Agricultural Sciences (SLU), Uppsala, Sweden.

Ulrike M. Irlich, ICLEI – Local Governments for Sustainability, Africa Secretariat, Century City, South Africa and Centre for Invasion Biology, Stellenbosch University, Matieland, South Africa.

Dave Kendal, School of Ecosystem and Forest Sciences, The University of Melbourne, Burnley, VIC, Australia.

Kalevi Korpela, School of Social Sciences and Humanities and Psychology, University of Tampere, Tampere, Finland.

Johannes Langemeyer, Institute of Environmental Science and Technology (ICTA), Universitat Autònoma de Barcelona, and Hospital del Mar Medical Research Institute (IMIM), Barcelona, Spain.

Kate E. Lee, School of Ecosystem and Forest Sciences and Melbourne School of Psychological Sciences, The University of Melbourne, Parkville, VIC, Australia.

Brenda B. Lin, Commonwealth Scientific and Industrial Research Organisation (CSIRO), Aspendale, VIC, Australia.

J. Scott MacIvor, Department of Biological Science, University of Toronto Scarborough, Toronto, Canada.

Adriano Martinoli, Department of Theoretical and Applied Sciences, Università degli Studi dell'Insubria, Varese, Italy.

Emily S. Minor, University of Illinois, Chicago, IL, USA.

Marcelo D. Miranda, Department of Ecosystems and Environment, Center for Applied Ecology and Sustainability (CAPES), Faculty of Agronomy and Forest Engineering, Pontificia Universidad Catolica de Chile, Santiago, Chile.

Jari Niemelä, Department of Environmental Sciences, University of Helsinki, Helsinki, Finland.

Alessandro Ossola, Centre for Smart Green Cities, Department of Biological Sciences, Macquarie University, Sydney, Australia. School of Ecosystem and Forest Science, The University of Melbourne, Richmond, Australia.

Sonia Reyes Paecke, Faculty of Agronomy and Forest Engineering, Center for Sustainable Urban Development (CEDEUS), Pontificia Universidad Catolica de Chile, Santiago, Chile.

Tytti Pasanen, School of Social Sciences and Humanities/Psychology, University of Tampere, Tampere, Finland.

Max R. Piana, Rutgers – The State University of New Jersey, New Brunswick, NJ, USA.

Richard Pouyat, U.S. Department of Agriculture Forest Service, Research & Development, Washington, DC, USA.

Clara C. Pregitzer, Natural Resources Group, New York City Department of Parks and Recreation, New York City, NY, USA. School of Forestry and Environmental Studies, Yale University, New Haven, CT, USA.

Eleanor Ratcliffe, School of Social Sciences and Humanities/Psychology, University of Tampere, Tampere, Finland.

Assaf Shwartz, Faculty of Architecture and Town Planning, Department of Landscape Architecture, Technion – Israel Institute of Technology, Haifa, Israel.

Stefan J. Siebert, Unit for Environmental Sciences and Management, North-West University, Potchefstroom, South Africa.

Alexis Dyan Smith, University of Illinois, Chicago, IL, USA.

Katalin Szlavecz, Department of Earth and Planetary Sciences, Johns Hopkins University, Baltimore, MD, USA.

Caragh G. Threlfall, School of Ecosystem and Forest Sciences, The University of Melbourne, VIC, Australia.

Lucas A. Wauters, Department of Theoretical and Applied Sciences, Università degli Studi dell'Insubria, Varese, Italy.

Ian Yesilonis, U.S. Department of Agriculture Forest Service, c/o Baltimore Ecosystem Study, Baltimore, MD, USA.

FOREWORD

We are all fascinated, attracted and inspired by nature. Even in today's modern world, for as urbanized and digital as we are, we cannot do without keeping in touch with nature. Many of the most loved children's cartoons are inspired by animals; we give flowers to our fiancées; we live closely with pet animals; we fill our homes with plants; our blood pressure drops when we enter an urban park from a busy city road; and to relax patients surgery theaters exhibit large pictures of forests, lakes and mountains on their walls.

We need nature. Not only because it gives us clean air, pure water, food and a stable climate among many other things, but because it has been our home for the vast majority of human history. The few decades of our 'modern' lifestyle are but a blink of an eye, compared to the over 2 million years of our existence on this planet . . . deeply immersed in nature. We need nature's presence around us; fascination for nature is literally in our genes.

However, our modern lifestyle is separating us more and more from the wild natural world. With over half of the world population already crammed into urban areas, many of the early experiences our children have with nature and biodiversity are in a city environment. At the same time, the relatively safe environment of cities is increasingly attracting wildlife from the surrounding natural habitats, offering the potential of magnificent wild encounters in an urban context.

In winter up to two million starlings flock in stereoscopic fashion over Rome's ancient ruins; peregrine falcons nest in Manhattan's skyscrapers in higher numbers than in most pristine mountains; large numbers of enormous fruit bats choose to roost in the trees of Colombo's city squares; marmoset monkeys thrive in the park at the foot of Rio de Janeiro's Sugarloaf Mountain; foxes and badgers settle in many European cities . . . not to mention the incredible leopards of Mumbai's Sanjay Gandhi park. And, finally, add the innumerable species such as butterflies, insects and birds to these wildlife celebrities, which make cities their home . . . and all of

a sudden the 'wild side' has conquered city parks, rivers, lakes and seaside prom-
enades, making them excellent environmental education open air laboratories.

We can also easily promote the rewilding of our cities: from 'wildlife gardening', using the right plants to attract all sorts of butterflies; the right berries to attract
birds; the right trees to create mini forest habitats perfect for owls and woodpeckers; to building the right artificial nests to attract innumerable birds, bats, insects
and much more. It's fun, it's exciting, it's incredibly fulfilling . . . and educational.
It offers everyone the opportunity to experience wildlife, take action and feel the
concrete and direct results.

It also promotes the understanding and appreciation of the incredible diversity
of life around us—something so easily ignored if we don't pay attention.

Urban biodiversity can be considered the doorway to wildlife and wild nature,
and a powerful vehicle for developing the awareness and empathy we desperately
need in a world where wild nature is shrinking and wildlife populations all around
the globe are declining steeply.

We will never save what we do not understand and love. Yes, nature conservation today is a complex business that involves the way we produce and consume,
manage markets and financial flows, and how we generate our energy, all with
implications for natural systems. But the bottom line is still intrinsically about the
value of nature to us and the emotions it evokes.

Big journeys start with small steps . . . and powerful experiences. And this book
is one of them. The more people are able to learn and experience nature and biodiversity, even in the simplest and most urbanized form, the more chance we have to
build a community of practice that supports the change we need to build a future
where people and nature are, finally, able to coexist in harmony.

Cities, designed to keep us apart from wild nature, can in fact turn out to be
our greatest allies for environmental education and conservation. We need to open
and foster a dialogue between research and practice to preserve urban biodiversity
in humanity's future.

Marco Lambertini
Director General
WWF International

PREFACE

Biodiversity – or biological diversity – represents the variety of living things around us. Genomes within organisms, species, habitats, and even whole ecosystems, all define what biodiversity is within a particular area. In the last decades, biodiversity has been at the centre of the research agenda due to the increasing challenges that many organisms and ecosystems face in adapting to environmental and climatic change.

Cities and towns represent the ultimate frontier of this scientific journey. Since researchers directed their attention to urban biodiversity, they rapidly realised its vast richness, the importance for the ecological functioning of cities and values for people living therein. In many cities worldwide, researchers discovered a diversity of species comparable to that found in some pristine natural ecosystems. They found that diverse urban habitats and communities of organisms help to maintain and regulate critical ecological processes which then sustain urban liveability and resilience. Researchers also revealed clear links between human well-being and urban biodiversity.

Regrettably, this body of knowledge remains too often secluded in academic publications only accessible to the community of researchers. New discoveries and information are largely unavailable to the community of practice, which could greatly benefit from them to improve current and future urban management, planning and design interventions. The community of practice itself consists of a large interdisciplinary group of professionals, students and even amateurs. City officials, urban managers, park managers, catchment managers, landscape architects, ecologists, politicians, engineers and designers are only some actors of the large community directly and indirectly dealing with urban biodiversity.

All the differences in priorities, languages and goals between researchers and the community of practice risk hindering the effective translation of scientific knowledge into policies, plans or best-management practices. This book provides the first contribution specifically designed to bridge the gap between research and practice by focusing on urban biodiversity. The idea for this book came from countless conversations between researchers and practitioners about the importance and the impact of research on urban biodiversity in the *real urban world*. We are truly grateful to all the people who provided suggestions, comments and help during the last three years. Staff of the ICLEI Urban Biodiversity Center in Cape Town (South Africa) were instrumental in the success of the preliminary survey conducted to scope knowledge and expectations about urban biodiversity among urban practitioners worldwide. We thank Shela Patrickson, Georgina Avlonitis, Elizabeth Metcalfe, Kobie Brand and Meggan Spires for their support and advice. Critical thoughts and encouragement were provided by members of the Urban Biodiversity, Ecology and Conservation (UrBEC) research group and the Green Infrastructure research group (GIRG) at the University of Melbourne, Australia, as well as the Urban Ecology research group at the University of Helsinki, Finland. Tim Hardwick, Amy Johnston and Ashley Wright (née Irons) at Routledge navigated us through this editorial adventure with their expertise and passion.

It is our hope that this book will be the beginning of a profitable *conversation* between researchers and community of practice on how to conserve *our* urban biodiversity for the generations to come.

Alessandro Ossola and Jari Niemelä
June 2017

1

BRINGING URBAN BIODIVERSITY RESEARCH INTO PRACTICE

Alessandro Ossola, Ulrike M. Irlich and Jari Niemelä

Cities and towns are the largest construct ever made on earth by a single species, *Homo sapiens*. Since humans abandoned their nomadic lifestyle, they started to build villages and towns, organising their communities and societies within and around them. Over time, urban expansion has caused numerous species, habitats and even whole ecosystems to be destroyed and altered. Peri-urban areas have been progressively cleared to make space for new developments and agricultural fields to feed an increasingly large human population. For the first time in biological history, novel urban habitats and ecosystems with no counterpart in the *natural world* have been created by human intervention, creativity and power (Gilbert 1989). Overall, this complex and rich variety of urban species, habitats and ecosystems is what we now commonly define as *urban biodiversity* (i.e. urban biological diversity).

Urban biodiversity is an ecological term frequently confused and interchangeably used with *urban green space* and *urban green infrastructure*, which principally derive from disciplines such as urban planning, landscape architecture and environmental management. By definition, urban biodiversity is a more general term that comprises the two other terms. For example, an urban stream is an aquatic ecosystem but not an urban green space, while an urban park is both an urban green space and an ecosystem. In writing this book, we have therefore privileged the more general term urban biodiversity, though the terms urban green space and green infrastructure are also used in numerous instances.

A brief historical overview of urban biodiversity research

Humans have examined urban biodiversity since the foundation of their early settlements. Much of the knowledge on biodiversity was pivotal for human livelihood, such as the provision of food and natural resources or the control of urban pests and diseases. Knowledge was mostly transmitted orally, though various manuscripts on urban animals and plants have been written from different civilisations over the centuries (Sukopp 2008). It was not until the onset of the twentieth century that a formal and more structured interest in urban biodiversity sparkled among researchers. Paradoxically, humans were the first species studied in an extensive urban research program. In fact, by applying early ecological principles, researchers at the Chicago School conducted numerous studies looking at relationships between human behaviour, social structures and urban environment during the 1920s and 1930s (Weiland and Richter 2012).

A further impulse to the research on urban biodiversity literarily germinated from the rubbles of World War II. Plant communities in degraded urban areas were in fact extensively studied by researchers of the Berlin School of Urban Ecology during the 1970s. Since then, numerous other studies, mostly focusing on landscape pattern and distribution of urban plants and animals, have been published (Figure 1.1). Researchers have found that the diversity of urban species is often comparable to that of peri-urban landscapes (Pickett *et al.* 2008), with numerous rare and threatened species still being discovered in urban areas nowadays (Ives *et al.* 2016). Even urban soils and waters, commonly believed to be biologically depleted, can host bountiful species (Szlavecz *et al.*, Chapter 2). In addition to the *native species* surviving in cities (i.e. original of that area), countless new species (i.e. *introduced* or *exotic species*) have been involuntarily and voluntarily introduced due to their economic and aesthetic value (Aronson *et al.*, Chapter 7). Over time, introductions have increased the species pool in various cities and towns worldwide. However, researchers found that introduced species can often be a significant threat to native species, particularly when these become invasive or pests (Wauters and Martinoli, Chapter 6; Hochuli and Threlfall, Chapter 4), potentially leading to the homogenisation of urban biological communities worldwide (McKinney 2006; Kowarik 2011).

It is not until after the first Convention on Biological Diversity (CBD) held in Rio de Janeiro, Brazil in 1992 (www.cbd.int/) that the term *urban biodiversity* appeared in the academic literature for the first time. Since then, the number of publications on urban biodiversity has experienced an exponential growth, particularly from the late 1990s (Figure 1.1).

In 1997, the establishment of two urban Long-Term Ecological Research (LTER) projects in Baltimore and Phoenix (USA) signed the beginning of the modern high-tech investigation of urban ecosystems that is still well underway today. Recent publications are starting to shed light on the important role of urban species and ecosystems in regulating key ecological processes and functions, such as soil nutrient cycling, urban heat island reduction and pollution mitigation (Pickett *et al.* 2008; Ossola *et al.* 2016). Almost a century since the Chicago School, humans also returned at the centre of the research agenda in the 2000s, recognised as critical

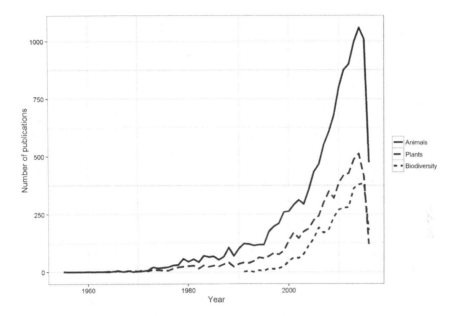

FIGURE 1.1 Number of academic publications published globally each year since 1955 having in their title, abstract or keywords the terms *urban* and *animals* (solid line), *urban* and *plants* (dashed line) and *urban* and *biodiversity* (dotted line) as censed in the Scopus database (Elsevier) on 7 June 2016 (www. scopus.com/). Publications for urban *animals* or *plants* concurrently containing the term *biodiversity* were excluded from graphing.

components of urban socio-ecological systems. Researchers discovered that humans can not only affect urban species and ecosystems, but also receive numerous benefits from them as *ecosystem services* (Langemeyer and Gómez-Baggethun, Chapter 3). For example, it has been shown that the vicinity to green spaces is positively reflected in real estate prices in numerous cities worldwide (e.g. Tyrväinen and Miettinen 2000). Urban green spaces are important for human cultural and recreational activities (e.g. festivals, sports), as well as for local food production (Lin and Egerer, Chapter 5). This can enhance the overall well-being of communities, but also offer new employment opportunities, besides those traditionally related to the management and maintenance of urban green spaces (e.g. gardeners, horticulturalists, arborists, etc.). Research suggests that it is ecologically and economically sound for cities to maintain and further develop their green space networks and green infrastructures (Minor *et al.*, Chapter 12) while planning for (Borgström, Chapter 10) and managing the overall urban biodiversity (Cilliers *et al.*, Chapter 11). The incorporation of people preferences, perceptions and values into biodiversity-sensitive urban designs can further facilitate the conservation of urban biodiversity (Lee and Kendal, Chapter 8; Ignatieva, Chapter 14). Importantly, researchers have found numerous links between urban biodiversity, and human physical and mental health in cities

(Korpela *et al.*, Chapter 9). For instance, visiting and spending time in green spaces has been shown to significantly reduce blood pressure. Even seeing green spaces from a window can significantly lower recovery time of hospitalised patients. Recent research confirmed that biodiversity in the surrounding landscape increases the diversity of bacteria on human skin, consequently lowering the incidence of allergies (Hanski *et al.* 2012). In addition to the direct human health benefits, enhancing green infrastructure of cities can indirectly facilitate innovative nature-based health solutions, such as green care (i.e. care of people using green infrastructure), and preventative medicine. A further recent avenue of research is exploring the interlinkages between urban biodiversity and cultural diversity, following the so-called *biocultural diversity concept* (e.g. Pretty *et al.* 2009; Maffi and Woodley 2010). Urban green spaces have a key role in maintaining not only urban biodiversity, but also the *socio-ecological memory* of urban systems (Barthel *et al.* 2010), thus sustaining important socio-ecological functions, resilience and adaptive capacity.

The beginning of the twenty-first century has seen the number of humans living in urban areas surpassing those living in rural areas (United Nations 2015), an unprecedented event in our evolutionary journey. We are approaching apace what researchers have defined the Anthropocene, a new geological epoch where human impacts have been manifested into the *earth's skin* – the layers of soil, sediment and ice across the globe (Waters *et al.* 2016). Much of the environmental and climatic change leading to this new epoch has been directly or indirectly generated from urban areas, anticipating unparalleled threats coming from rampant and seemingly unstoppable urban futures. Many cities have been built near natural biodiversity hot-spots, and future urban expansion poses new, challenging scenarios not only for urban species, habitat and ecosystems, but also for those that persist beyond urban fringes (Seto *et al.* 2012). Despite the proximity of many cities and towns to pristine ecosystems, humans are becoming progressively detached from nature, spending more time in digital rather than outdoor worlds, with potential deleterious effects for biological conservation (Shwartz, Chapter 13).

This represents an important and stimulating time for urban socio-ecological research and the thousands of researchers that will continue to tackle these numerous issues on urban biodiversity in the coming years (McDonnell and MacGregor-Fors 2016).

BOX 1.1

Since the early 1990s, more than 3500 publications specifically related to *urban biodiversity* have been published in the academic literature (based on the Scopus database (Elsevier), 7 June 2016). However, this represents only a small fraction of all the publications related to urban biodiversity. Many others, for example on urban animals and plants, do not explicitly mention the term *urban biodiversity* in their texts (Figure 1.1). Most publications on urban biodiversity (88 per cent) are represented by articles published in academic journals, followed by conference papers (6 per cent), books and chapters (5 per cent) and other types of documents (1 per cent). In particular, articles have been published in more than 150 different academic journals that cover a vast range of disciplines such as ecology, planning, environmental management, engineering, sociology, policy, conservation, horticulture, toxicology, sustainability, hydrology, forestry and economics. This demonstrates that, particularly in the last 25 years, urban biodiversity has become a vast research topic able to transcend environmental disciplines and stimulate the interest of researchers from numerous other disciplines. The latest research on urban biodiversity has also become more geographically widespread, with researchers from 131 countries having published at least one contribution on the topic to date. Overall, most of the research on urban biodiversity has focused on cities across Europe, Northern America and Australia (Figure 1.2). Urban biodiversity in many countries in Central and Southern America, Africa and Asia has been proportionally less studied (Figure 1.2). This represents a clear gap in the academic literature that future research should timely address, some of these developing countries being located in biodiversity hot-spots (Seto *et al.* 2012), and in the epicentre of current and future urbanisation (United Nations 2015).

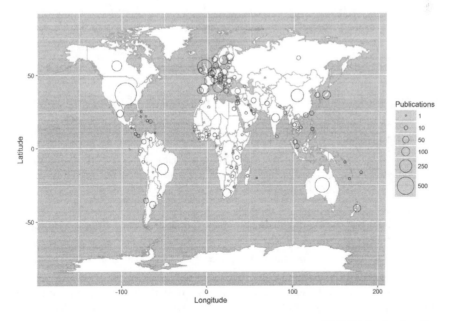

FIGURE 1.2 *(continued)*

(continued)

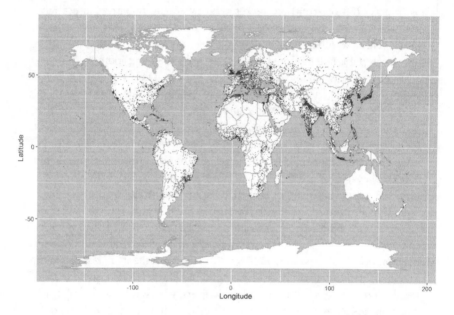

FIGURE 1.2 (Upper panel) Map of the world's cities with each point representing a city with population greater than ~40.000 inhabitants in 2006 (drawn with the R package *maps*, Brownrigg 2016). (Lower panel) Map of the global distribution of academic publications published in each country since 1992 having the terms *urban* and *biodiversity* in their title, abstract or keywords as censed in the Scopus database (Elsevier) on 7 June 2016 (www.scopus.com/).

Urban biodiversity and the global community of practice

In a rapidly urbanising world, subnational and local governments play an increasingly vital role in protecting and managing urban biodiversity. Local governments are in charge of providing many services to human settlements (e.g. drinking water, waste removal and open space management), which may have negative or positive impacts on species and ecosystems. Hence, governments play a critical role in managing demand for local ecosystem services and natural resources, in contributing to and implementing strategies that enable the protection of these. Increasingly, strategies are framed by international agreements that set shared and common goals, such as those related to the Aichi Targets and the National Biodiversity Strategy and Action Plans (NBSAPs). These goals are then often implemented by subnational and local governments that are well placed in taking on the role of urban biodiversity management and protection due to their detailed knowledge of the urban systems they are operating in (Figure 1.3).

In the last three decades, ICLEI – Local Governments for Sustainability (ICLEI), an international association of government associations, has been the key driver in advocating for the sustainable management of natural resources at the global level. Through working in close partnership with the Secretariat of the Convention on Biological Diversity (SCBD) and other instrumental partners (IUCN, UN-Habitat, Stockholm Resilience Centre), ICLEI has supported local governments in their endeavour to manage urban biodiversity. Furthermore, through its projects and initiatives, ICLEI has provided a platform for local governments to voice their challenges, successes and objectives to the international community. ICLEI's flagship biodiversity programme (the Local Action for Biodiversity (LAB) Pioneer Programme) demonstrated the importance of local governments' contributions to natural resource management and to the successful implementation of the CBD.

FIGURE 1.3   Biodiversity management should be vested with subnational and local government as a result of six key factors. Reproduced and adapted with permission from ICLEI/SCBD 2017, www.cbd.int/subnational / draft-guidelines.

The LAB Programme culminated in the adoption of Decision X/22 by Parties to the CBD, at the 10th meeting of the Conference of the Parties (COP10) in Nagoya, Japan, October 2010 (www.cbd.int/doc/decisions/cop-10/cop-10-dec-22-en.pdf). This decision approved the Plan of Action on Sub-national Governments, Cities and Other Local Authorities for Biodiversity (2011–2020), which provides guidance to the Parties on how to support local governments towards collaborative implementation of the Convention. The Cities and Biodiversity Outlook (SCBD 2012) assessed the links and opportunities between urbanisation and biodiversity, to support this decision.

A global partnership was initiated, bringing key stakeholders, networks and initiatives together, with the aim of supporting local and subnational action for biodiversity (www.cbd.int/subnational/partners-and-initiatives). Since 2008, the Global Biodiversity Summit of Cities and Subnational Governments has been held biannually in parallel with the CBD COPs, providing an important platform for engaging local and subnational governments in global biodiversity conservation actions.

Local and subnational governments can pledge to protect biodiversity and recognise its importance, through the signing of commitments, such as the *Durban Commitment* (http://archive.iclei.org/). Similarly, the *Quintana Roo Communiqué on Mainstreaming Local and Subnational Biodiversity Action* (www.cbd.int/cop/cop-13/other/pr-quintana-roo-en.pdf) was endorsed at the 5th Global Biodiversity Summit of Cities and Subnational Governments in Mexico in 2016, further stimulating urban biodiversity conservation initiatives.

Cities and towns managing urban biodiversity

Over the last two decades, numerous local municipalities worldwide have started to take action to preserve urban biodiversity. For example, the European Capitals of Biodiversity run by Deutsche Umwelthilfe is a project that honours and highlights municipalities that are aware of, and committed to, urban biodiversity, and subsequently communicates their performance at both national and European levels (www.capital-biodiversity.eu).

Similarly, the partnership between the International Union for Conservation of Nature (IUCN) and ICLEI's LAB Programme, initiated in 2006, encouraged local governments to manage biodiversity. The programme followed a step wise process that was action oriented and customised for local and regional authorities around the world, seeking to improve biodiversity planning and management at the local level. A key focus of the programme was to assist local governments in the development of Local Biodiversity Strategy and Action Plans (LBSAPs) and to mainstream biodiversity into land-use and decision making processes. The development of guidelines for urban practitioners aimed at building capacity and empowering local authorities in their endeavour to manage urban biodiversity.

The LAB initiative has since expanded to include more cities and has branched into thematic streams, including LAB Climate Change and Biodiversity; LAB Wetlands and Communities; LAB Wetlands South Africa (http://cbc.iclei.org/programmes/lab/).

The outputs produced (e.g. guidelines and assessment reports) were not formally based on urban ecological research, though some consultation with scientists has occurred in the preparation of these.

Resources developed to assist cities and towns

Over time, several tools and guidelines have been developed to assist local governments in planning for biodiversity and mainstreaming these into land use planning processes. Tools developed are freely available (e.g. http://cbc.iclei.org/resources) and are constantly being built upon. ICLEI and the CDB, in partnership with several other partners have been instrumental in developing, distributing and building capacity around these resources to ensure the uptake at the local level. A subset of guidelines and tools are discussed below:

Guidelines

ICLEI in association with IUCN and the CBD published the *LAB Guidebook* (ICLEI – Local Governments for Sustainability 2010), a resource providing practitioners with guidelines on the various aspects of biodiversity management in the challenging urban context. This publication collates biodiversity best-management practices from around the world and contains practical advice, examples, and references for further information. It draws on a variety of sources with a focus on the rich experience of leading cities and regions from around the world, though these are not generally grounded in the scientific literature.

The *Local Biodiversity Strategy and Action Plan (LBSAP) Guidelines: an aid to municipal planning and biodiversity conservation* (ICLEI/UNU-IAS/SCBD 2010) are intended to assist local governments in developing their LBSAPs and assist with their biodiversity planning processes. The Guidelines provide background guidance and supporting information for planning at the local level.

The renowned study on *The Economics of Ecosystem and Biodiversity* (TEEB; www.teebweb.org/) draws attention to the economic benefits of biodiversity and emphasises the growing costs of biodiversity loss and ecosystem degradation. The TEEB's *Manual for Cities: Ecosystem Services and Urban Management* (TEEB 2011) aims to assist urban and regional policymakers and planners to assess the value of natural assets and consider opportunities and trade-offs in their policy and planning decisions. The manual offers an easily understandable introduction to ecosystem services. It then covers topics such as how to determine their value and how to incorporate ecosystem services into municipal functioning as a long-term investment to enhance existing municipal management.

Cities Biodiversity Index

The *Cities Biodiversity Index* (CBI) is currently the only index designed to measure biodiversity, ecological footprint and conservation efforts in cities to achieve the goals of the CBD. The index, designed through a consultative process leaning heavily on experts in the field, including practitioners and scientists. Technical experts that were involved in the development of the index included delegates from the Stockholm Resilience Centre, ICLEI, IUCN, London School of Economics and others. The CBI was developed as a self-assessment tool to be used by cities to determine the biodiversity status of a particular city and as a tool to monitor changes (Chan *et al.* 2010). While numerous cities have been using this tool, there is, however, no formal reporting mechanism in place to track the number of cities using this Index.

Cities implementing action plans and policies

The initial phase of the LAB project involved 21 pioneering cities in developing a synthesis report and LBSAPs (City Biodiversity Reports and LBSAPS can be found at http://cbc.iclei.org/resources). In preparation for COP 13, the CBD released a progress report (UNEP/CBD/COP 2016), indicating that 19 Parties had developed at least one subnational BSAP.

In 2015, an independent study was concluded that analysed LBSAPs (which included LAB cities LBSAPs, LBSAPs submitted to the CBD, and a range of other LBSAPs) (Pierce 2015). The study analysed a total of 65 plans and found a marked increase in the number of plans published after 2008, coinciding with the *CBD COP Decisions* (IX/28; IX/22 and XI/8 in particular) encouraging local and subnational action. Of the plans analysed, 48 plans were specific biodiversity plans while 17 other types were included, as they covered biodiversity aspects (e.g. Comprehensive Plans; Sustainability Plans; Climate Change Plans; Wetland Plans). A large number of plans were found to be from the developed world, with less than 20 per cent of the plans coming from developing countries (Table 1.1).

LBSAPs differ in structure and content across the globe because each municipality adapts their content to their needs. However, some of the general structure and content is outlined in the LBSAP Guidelines, which are widely used and · adapted in the development of these plans.

More importantly, however, are the lessons learned through the LAB process and through working with local governments on the translation of BSAPs into tangible projects and mainstreaming these documents into land-use planning and decision making processes.

To ensure political buy-in and subsequent implementation through unlocking municipal funds, LBSAPs need to be adopted through municipal processes. While the signing of commitments, such as the Durban Commitment, is generally a first step, subsequent action and mainstreaming is critical in ensuring the uptake of the biodiversity strategy and action plans in municipalities, and to ensure the implementation of these biodiversity strategy and action plans. Hence, LBSAPs should be mainstreamed into processes such as masterplans, development strategies and plans or the like, as well as policies, keeping in mind the urban development considerations. This can also result in cross-sectoral benefits through planning with nature, where urban blue–green infrastructure can be leveraged to support municipal service delivery (e.g. wetland management for water provisioning and flood attenuation).

TABLE 1.1 List of the 48 Local Biodiversity Strategy and Action Plan (LBSAPs) from 17 countries (Pierce 2015).

Country	Location	Plan type	Area	Date
Australia	Joondalup	LBSAP	City	2009
Australia	Melbourne	Conservation Plan	City	2015
Brazil	Curitiba	LBSAP	City	2012
Brazil	SaoPaulo	LBSAP	City	2011
Canada	Edmonton	LBSAP	City	2009
England	Birmingham & Black County	LBAP	Region	2004
England	Brighton & Hove	LBAP	City	2016
England	Bristol	LBAP	City	2008
England	Exeter	LBAP	City	2009
England	Greater Manchester	LBAP	County	2009
England	Greenwich (in London)	LBAP	Sub-city	2009
England	Kingston upon Hull	LBAP	City	2008
England	Leeds	LBAP	City	2000
England	Leicester	LBSAP	City	2011
England	Lincoln	LBAP	City	2006
England	London Region	LBAP	Region	2010
England	Newcastle & North Tyneside	LBAP	Region	2011
England	North Merseyside County	LBAP	Region	2008
England	Norwich	LBAP	City	2002
England	Portsmouth	LBAP	City	2012
England	Sheffield	LBAP	City	2002
England	Southampton	LBAP	City	2006
England	Westminster (in London)	LBAP	City	2008
England	Worcestershire	LBAP	City	2008
France	Paris	LBAP	City	2011
Germany	Bonn	LBSAP	City	2008
India	Sikkim	BAP	State	2016
Ireland	Belfast	LBAP	City	2007
Ireland	Cork City (in Cork County)	LBAP	City	2009
Ireland	Dublin	LBAP	City	2008
Ireland	Dun Laoghaire-Rathdown	LBAP	County	2009
Japan	Aichi Prefecture	LBSAP	State	2013
Japan	Chiba Prefecture	LBSAP	State	2012
Japan	Nagoya	LBSAP	City	2014
Japan	Saitama Prefecture	LBSAP	State	2012
Korea	Seoul	Biodiversity Report	City	2008
Mexico	Mexico City	LBSAP	City	2017
New Zealand	Auckland	Indigenous BSP	City	2016
New Zealand	Waitakere (now Auckland)	LBSAP	City	2008
Scotland	Edinburgh	LBAP	City	2010
Scotland	Glasgow	LBAP	City	2000
Singapore	Singapore	NBSAP	City-state	2009
South Africa	City of Cape Town	LBSAP	City	2009
South Africa	City of Johannesburg	LBSAP	City	2009
South Africa	eThekwini (Durban)	LBSAP	City	2008
United States	Schaumburg, IL	LBAP	City	2008
United States	Chicago, IL	LBAP	City	2011
Wales	Cardiff	LBAP	City	2008

Challenges faced by local authorities in implementing biodiversity plans

Local authorities are mandated to provide basic service delivery to communities. In a rapidly urbanising world, the demand for service delivery is rapidly increasing and local authorities are often under capacitated, in man power and budgets, as well as skill sets (Chukwudi 2015; Irlich *et al.* 2017). In the South African context, a recent paper unpacked the challenges faced by municipalities in the management of invasive species and makes some recommendations to overcome these related to public–private partnerships, capacity and knowledge building and funding (Irlich *et al.* 2017). While the paper focussed on South Africa, the challenges in other developing and developed countries are very similar in nature. In fact, some of the cities participating in the LAB project summarised their challenges as the lack of adequate financial resources, land ownership issues, poor coordination among different departments or bodies responsible for management, and urbanisation and development pressures. While the integration of scientific knowledge about urban biodiversity in international advocacy actions, policy making and planning has been rarely attempted, this could represent a logical and desirable step to guide future actions aimed at urban biodiversity conservation.

Connecting urban biodiversity research and practice

Robust knowledge on urban biodiversity is critical to improving current and future practices related to urban planning, design and management within modern urban governance structures. Research can guide cities and towns to maximise urban conservation outcomes and the relative provision of ecosystem services, while minimising the associated costs, uncertainties, trade-offs and risks (Pett *et al.* 2016). However, the translation of academic knowledge into effective practices is frequently seen as an expensive, time-consuming and unfeasible exercise. This is possibly caused by the disconnection between research and practice. Practitioners and researchers come from disparate professional worlds with different languages, expertise, priorities and timeframes. For instance, practitioners do not necessarily know how to approach an increasingly large body of academic literature and distil actionable information from scientific publications. On the other hand, researchers do not always understand the intricacies and constraints of planning and regulatory processes. It may be even challenging for some researchers to understand that urban planning is not a scientific discipline per se, but rather a social action within scientific, technological and legal bases (Nilsson and Florgård 2009). The community of practice dealing with urban biodiversity is often working in under capacitated departments constrained by numerous impediments set by politics, municipal processes and procurement.

Thus, the effective translation of urban biodiversity research into practice begins from the *reciprocal recognition and appraisal*, from both researchers and practitioners, of the differences and commonalities of the respective professions.

For researchers to be able to enhance the use of their findings in the urban practice, it is vital to understand where their intellectual inputs into management, planning and regulatory processes are feasible and how this can be effectively delivered (Yli-Pelkonen 2008). Each city and town is a unique blend of cultural, economic and socio-ecological characteristics and legacies that have been shaped over time by the environment and humans. Researchers, who often try to find *general trends and patterns* in the systems they study, need to understand and recognise the specificity of the cities and towns they work in. For example, strategies that have been proved effective for urban biodiversity conservation in one city might be inadequate or impractical in other cities characterised by different ecological and socio-economic context, size or cultural background. Thus, an appropriate *scaling* of more general research findings into local and particular urban contexts could lead to a faster and more profitable translation of urban biodiversity research into practice. Further, the capacity challenges that cities face highlight the importance of local authorities to lean on external experts and researchers to aid in urban biodiversity management and decision making processes. Cities should seek scientific advice for relevant matters by partnering with researchers where possible. This can be in the form of embedding researchers within the municipal structures or employing researchers directly; outsourcing and funding scientific research; partnering with universities to get students or research staff to address key research needs; linking up with organisations such as the Thriving Earth Exchange (http://thrivingearthexchange.org/); or joining research networks, such as the Urban Biodiversity Research Coordination Network (UrBioNet; http://urbionet.weebly.com/). UrBioNet is a global research network that brings together researchers and practitioners to address key questions on urban biodiversity. Another worldwide network, the URBIO network (URban BIOdiversity and Design; www.urbionetwork.org), provides a scientific network for education and research to promote the implementation of the CBD in urban areas. It aims to foster scientific exchange between researchers, practitioners and stakeholders through hosting international scientific conferences and workshops. Examples of such successful engagements include the City of Cape Town, South Africa, addressing research needs on restoration and invasive species related aspects (Irlich *et al.* 2017; Mostert *et al.* 2017).

Further, urban governance processes are rarely regulated through *distributed decision making*, but rather one or a few practitioners are usually in charge of particular actions and tasks based on existing legislative settings of a particular city or country. Thus, researchers could first engage with these key practitioners by establishing collaborations based on reciprocal trust to then amplify positive outcomes to the larger community of practice through *trickle down effects* and *behaviour mimicry*. Furthermore, there are often varying and competing interests involved in planning and regulatory processes, some of which are stronger than others (e.g. politics, economy). Although these interests may prevail over the socio-ecological ones, it is vital that researchers demonstrate how their research could provide *added values* to a multitude of processes, rather than single ones.

While innovative solutions for urban biodiversity conservation are frequently conceived and realised by practitioners, these could be designed with the aid of researchers able to test and validate their effectiveness (Shwartz *et al.* 2014). However, practitioners are often not aware that the profession of a researcher is a very particular one. Professional careers are intimately tied to the researcher's productivity in terms of number and quality of academic publications, the journals where these are published, the number of citations each publication received over time (i.e. impact), and the monetary resources a researcher is able to secure from public and private funding agencies. Thus, every attempt to invest time and resources towards community outreach or science communication has to be carefully balanced with research duties, teaching and administrative tasks.

Each researcher is often an expert in a particular sub-discipline or topic. Being increasingly nationally and internationally mobile, researchers with a particular expertise or knowledge might also work for a limited time in a city or a country. *Timing* is therefore a further critical aspect to consider, together with a clear operational understanding of the different funding schemes accessible to either researchers and/or the community of practice (e.g. timeframes, expectations, responsibilities, etc.). This is particularly important when research-practice collaborations are established through commonly designed experiments, projects and initiatives.

Although there are numerous gaps in our understanding of the dynamics of urban biodiversity, simply more research is not sufficient to enhance the practical use of academic knowledge. The challenges for scientists lies in improving the evidence base of urban biodiversity and socio-ecological systems so that we are able to better predict future changes and thereby provide guidance to urban practitioners (Niemelä *et al.* 2011). Addressing issues of relevance to the community of practice is of the utmost importance for the credibility and practical relevance of research (Colding 2011; Pauleit *et al.* 2011). To achieve this, a research approach integrating different scientific disciplines with the needs of the city planners, decision makers and residents is necessary (McPhearson *et al.* 2016).

Ahern *et al.* (2014) recently proposed a model for *safe-to-fail* adaptive urban design to provide a framework for integrating research, professional practice and stakeholder participation. The model is a transdisciplinary working method and includes experimental design guidelines, monitoring and assessment protocols, and strategies for delivering specific urban ecosystem services integral with urban development. The *safe-to-fail* adaptive urban design model encourages innovation in low-risk contexts and it could be well up-scaled and integrated with strategies involving new technologies, big data and citizen participation (Dobbs *et al.*, Chapter 15). In fact, citizen science has proven to be very successful in supporting biodiversity management and protection, with most of the LAB cities having active programmes to engage municipalities with the public and encourage their participation (see the Biodiversity Reports of Bonn (Germany), Calgary and Edmonton (Canada), Jerusalem (Israel) and São Paulo (Brazil) for some examples; http://cbc.iclei.org/resources/). Co-designing and co-producing scientific knowledge can represent a challenging yet promising opportunity for all of us, researchers and practitioners, to better preserve the world's urban biodiversity and secure a vital future.

References

Ahern J., Cilliers S. and Niemelä J. (2014) The concept of ecosystem services in adaptive urban planning and design: A framework for supporting innovation. *Landscape and Urban Planning* 125: 254–259.

Barthel S., Folke C. and Colding, J. (2010) Social-ecological memory in urban gardens: Retaining the capacity for management of ecosystem services. *Global Environmental Change* 20: 255–265.

Brownrigg R. (2016) *Maps: Draw Geographic Maps*. https://cran.r-project.org/package=maps. R package version 3.1.0.

Chan L., Calcaterra E., Elmqvist T., Hillel O., Holman N., Mader A. and Werner P. (2010) *User's Manual for the City Biodiversity Index*. Latest Version: 27 September 2010 www.cbd. int/authorities/doc/User%27s%20Manual-for-the-City-Biodiversity-Index27Sept2010. pdf (accessed 16 February 2017).

Chukwudi A.S. (2015) Manpower development, capacity building and service delivery in Ife-East Local Government Area, Osun State, Nigeria. *Journal of Public Administration and Policy Research* 7(1): 1–14.

Colding J. (2011) The role of ecosystem services in contemporary urban planning. In: Niemelä J., Breuste J., Elmqvist T., Guntenspergen G., James P. and McIntyre N. (eds). *Urban Ecology: Patterns, Processes, and Applications*, pp. 228–237. Oxford University Press, Oxford.

Gilbert O.L. (1989) *The Ecology of Urban Habitats*. Chapman & Hall, London.

Hanski I., von Hertzen L., Fyhrquist N., Koskinen K., Torppa K., Laatikainen T., Karisola P., Auvinen P., Paulin L., Mäkelä M.J., Vartiainen E., Kosunen T.U., Alenius H. and Haahtela T. (2012) Environmental biodiversity, human microbiota, and allergy are interrelated. *Proceedings of the National Academy of Sciences* 109: 8334–8339.

ICLEI – Local Governments for Sustainability (2010) *Local Action for Biodiversity Guidebook: Biodiversity Management for Local Governments*. Laros M.T. and Jones F.E. (eds).

Irlich U.M., Potgieter L., Stafford L. and Gaertner M. (2017) Recommendations for municipalities to become compliant with national legislation on biological invasions. *Bothalia* 47(2): a2156.

Ives C.D., Lentini P.E., Threlfall C.G., Ikin K., Shanahan D.F., Garrard G.E., Bekessy S.A., Fuller R.A., Mumaw L., Rayner L., Rowe R., Valentine L.E., Kendal D. (2016) Cities are hotspots for threatened species. *Global Ecology and Biogeography* 25: 117–126.

Kowarik I. (2011) Novel urban ecosystems, biodiversity, and conservation. *Environmental Pollution* 159(8–9): 1974–1983.

Maffi, L. and Woodley, E. (2010) *Biocultural Diversity Conservation: A Global Sourcebook*. Routledge, London.

McDonnell M.J. and MacGregor-Fors I. (2016) The ecological future of cities. *Science* 352: 936–938.

McKinney M.L. (2006) Urbanization as a major cause of biotic homogenization. *Biological Conservation* 127: 247–260.

McPhearson T., Pickett S.T.A., Grimm N., Niemelä J., Alberti M., Elmqvist T., Weber C., Haase D., Breuste J. and Qureshi S. (2016) Advancing urban ecology toward a science of cities. *BioScience* 66(3): 198–212.

Mostert E., Gaertner M., Holmes P.M., Rebelo A.G. and Richardson D.M. (2017) Impacts of invasive alien trees on threatened lowland vegetation types in the Cape Floristic Region, South Africa. *South African Journal of Botany* 108: 209–222.

Niemelä J., Breuste J., Elmqvist T., Guntenspergen G., James P. and McIntyre N. (eds) (2011) *Urban Ecology: Patterns, Processes, and Applications.* Oxford University Press, Oxford.

Nilsson K.L. and Florgård C. (2009) Ecological scientific knowledge in urban and land-use planning. In McDonnell M.J., Hahs A.K. and Breuste J.H. (eds), *Ecology of Cities and Towns: A Comparative Approach*, pp. 549–556. Cambridge University Press, Cambridge.

Pauleit S., Liu L., Ahern J. and Kazmierczak A. (2011) Multifunctional green infrastructure planning to promote ecological services in the city. In Niemelä J., Breuste J., Elmqvist T., Guntenspergen G., James P. and McIntyre N. (eds), *Urban Ecology: Patterns, Processes, and Applications*, pp. 272–285. Oxford University Press, Oxford.

Pett T.J., Shwartz A., Irvine K.N., Dallimer M. and Davies Z.G. (2016) Unpacking the people–biodiversity paradox: A conceptual framework. *BioScience* 66, 576–583.

Pickett S.T.A., Cadenasso M.L., Grove J.M., Groffman P.M., Band L.E., Boone C.G., Burch W.R.jr., Grimmond C.S.B., Hom J., Jenkins J.C., Law N.L., Nilon C.H., Pouyat R.V., Szlavecz K., Warren P.S. and Wilson M.A. (2008) Beyond urban legends: An emerging framework of urban ecology, as illustrated by the Baltimore Ecosystem Study. *BioScience* 58, 139–150.

Pierce J.R. (2015) Planning for urban biodiversity in a divided world. Masters Thesis, Cornell University, USA.

Pretty J., Adams B., Berkes F., de Athayde S.F., Dudley N., Hunn E., Maffi L., Milton K., Rapport D., Robbins P., Sterling E., Stolton S., Tsing A., Vintinnerk E. and Pilgrim S. (2009) The intersections of biological diversity and cultural diversity: Towards integration. *Conservation & Society* 7: 100–112.

Ossola A., Hahs A.K., Nash M.A. and Livesley S.J. (2016) Habitat complexity enhances comminution and decomposition processes in urban ecosystems. *Ecosystems* 19: 927–941.

SCBD (2012) *Cities and Biodiversity Outlook.* Secretariat of the Convention on Biological Diversity. Montreal, CA.

Shwartz A., Turbé A., Julliard R., Simon L. and Prévot A.-C. (2014) Outstanding challenges for urban conservation research and action. *Global Environmental Change* 28: 39–49.

Seto K.C., Güneralp B. and Hutyra L.R. (2012) Global forecasts of urban expansion to 2030 and direct impacts on biodiversity and carbon pools. *Proceedings of the National Academy of Sciences of the United States of America* 109(40): 16083–16088.

Sukopp H. (2008) On the early history of urban ecology in Europe. In Marzluff J.M., Shulenberger E., Endlicher W., Alberti M., Bradley G., Ryan C., Simon U. and ZumBrunnen C. (eds), *Urban Ecology: An International Perspective on the Interaction Between Humans and Nature*, pp. 79–97. Springer US.

TEEB (2011) *TEEB Manual for Cities: Ecosystem Services in Urban Management.* Available at www.teebweb.org.

Tyrväinen L. and Miettinen A. (2000) Property prices and urban forest amenities. *Journal of Environmental Economics and Management* 39: 205–223.

United Nations (2015) *World Population Prospects: The 2015 Revision, Data Booklet*, United Nations, Department of Economic and Social Affairs, Population Division ST/ESA/SER.A/377. Available at http://esa.un.org/unpd/wpp/Publications/Files/WPP2015_DataBooklet.pdf (accessed 6 June 2016).

UNEP/CBD/COP (2016) Update on the Progress in Revising/Updating and Implementing National Biodiversity Strategies and Action Plans, including National Target. UNEP/CBD/COP/13/8/Add.1/Rev.1. Available at www.cbd.int/doc/meetings/cop/cop-13/official/cop-13-08-add1-rev1-en.doc (accessed 1 August 2017).

Waters C.N., Zalasiewicz J., Summerhayes C., Barnosky A.D., Poirier C., Gałuszka A., Cearreta A., Edgeworth M., Ellis E.C., Ellis M., Jeandel C., Leinfelder R., McNeill J.R., Richter D.deB., Steffen W., Syvitski J., Vidas D., Wagreich M., Williams M., Zhisheng A., Grinevald J., Odada E., Oreskes N. and Wolfe A.P. (2016) The Anthropocene is functionally and stratigraphically distinct from the Holocene. *Science* 351: 6269.

Weiland U. and Richter M. (2012) *Applied Urban Ecology: A Global Framework*. Blackwell Publishing, Chichester, UK.

Yli-Pelkonen V. (2008) Ecological information in the political decision making of urban land-use planning. *Journal of Environmental Planning and Management* 51: 345–362.

2

SOIL AS A FOUNDATION TO URBAN BIODIVERSITY

Katalin Szlavecz, Ian Yesilonis and Richard Pouyat

Introduction

As cities grow and spread over the globe so do urban soils. This brown infrastructure provides important ecosystem services such as a medium for vegetation growth, a system for water supply and purification, a recycling system for nutrients and organic wastes, and a habitat for soil organisms. The functions of these soils, in an urban environment, are affected by both physical and chemical processes; for example, the physical grading of the soil for building construction and nitrogen deposition to the surface of the soil, respectively. These processes directly and indirectly affect soil biota and the function of the soil ecosystem. In this chapter, we will discuss physical and chemical processes commonly affecting soils in urban areas and the interaction between the living and non-living components of the soil ecosystem. We highlight the importance of a diverse, healthy soil ecosystem and offer some ideas to consider when managing for soil health. Soils are a fundamental natural resource that needs to be conserved for the health of the environment and humans.

Natural and anthropogenic soil forming factors

Soil is defined as "the unconsolidated mineral or organic matter on the surface of the earth that has been subjected to and shows effects of genetic and environmental factors of: climate (including water and temperature effects), and macro- and microorganisms, conditioned by relief, acting on parent material over a period of time" (SSSA 2017). Under natural conditions, soil development (pedogenesis) is a slow process that takes hundreds to thousands of years to occur driven by five soil forming factors: parent material (geologic rock and organic materials), climate (temperature and precipitation), biota (vegetation, microbes, and soil animals), topography (slope, aspect, and landscape position), and time (Jenny 1941). Soil properties are formed through four basic processes: additions, transformations, translocations and losses. Under similar environmental conditions, parent material, and vegetation, soils develop with similar characteristic horizons. Soil development is long-term process; for example, it takes approximately 4,000 years for the uppermost weathered part of the soil profile of a Gray-Brown Podzolic (Hapludalf) soil to form from weathered loess in Iowa, USA (Arnold and Riecken 1964). Naturally forming soils have certain physical and chemical features that allow them to provide ecosystem services such as a medium for vegetation growth, a system for water supply and purification, a recycling system for nutrients and organic wastes, and a habitat for soil organisms.

In an undisturbed environment, managers can predict soil properties based on the relative importance of these soil forming factors (Lin 2011). For example, the ridge or hilltop generally accumulates organic matter while on the slope there is less organic matter due to erosion and to lack of water infiltration. At the bottom of the hill, the soils tend to be high in both organic matter and moisture. In general, this predictability allows managers to be informed when making decisions on land use and the management to maximize productivity.

In cities human activity becomes a major force in soil evolution. The landscape is divided and determined by ownership, and within these new boundaries, past and present human actions, in addition to natural soil forming factors, define the characteristic of soil (Pouyat *et al.* 2010; Ossola and Livesley 2016). Soils in the urban

FIGURE 2.1 Urban soils are diverse due to a combination of natural and anthropogenic soil forming factors, site history, and management. Soil horizons (O, A, B C) under remnant patches of the native vegetation are usually retained or less modified. Even under the same land cover (e.g. grass) the vertical profile can be quite different. Organic matter (OM) can be buried under different kinds of fill material. Source: Ian Yesilonis, unpublished.

environment are affected by the degree of disturbance, past history, current land use, land cover, and management, all of which create a heterogeneous template (Figure 2.1). Usually, the term "urban soil" refers to human-altered and transported soils within the urban environment (soil profiles on the right side of Figure 2.1). It is important to recognize that such human influenced soils are found in many other places such as industrial, traffic, mining, and military areas, often abridged as SUITMA (Morel et al. 2015). Additionally, soils that have more or less retained their original profile, usually under remnant patches of the local biome, can also be found within urban areas (Figure 2.1). These soils may be less disturbed physically, but still affected by other anthropogenic factors such as pollution and invasive species. In this chapter, urban soil refers to all soils found within urban areas.

Properties of urban soils

Physical properties

In urban landscapes, soils are often disturbed and managed, eventually leading to alteration or loss of the natural soil profile. A variety of physical disturbances affect soil composition and function. The International Committee on Anthropogenic Soils classifies these disturbances into the following categories: accelerated erosion, land filling, land leveling, surface removal, contamination, sedimentation, deep plowing and logging, severe compaction by machinery, and artificial saturation (ICOMANTH 1997). For example, if a site is being developed for a building such as a house or store, the land must be leveled prior to construction. Heavy machinery grades the landscape which includes surface removal, land filling, land leveling, and compaction. The fill used to level the landscape may be some mixture of natural soils, construction debris, material dredged from waterways, or coal ash (Figure 2.1). The mixing of the natural materials with different types of fill can be only a couple of centimeters thin, or several meters deep. Furthermore, during construction, the top soil is usually removed in the process of preparing the land for building structures. Chen et al. (2013) have shown that scraping topsoil, stockpiling it for approximately one month, and replacing the topsoil of construction sites resulted in a loss of more than a third of soil organic carbon from the topsoil. This loss of soil organic carbon negatively affects soil quality and function. For example it is estimated that soil organic matter (SOM) can hold up to 20 times its weight in water (Reicosky 2005). Loss of topsoil and SOM results in diminished water holding capacity and affects infiltration. Altered water movement through soil profiles are unpredictable and thus challenging to manage.

In urban environments most of the surfaces are covered by asphalt, concrete, roofing, or some other impervious layer. Covering the soil with an impervious layer, or soil sealing, is one of the most drastic changes in urban areas. It, effectively disrupts the natural exchange of water and gases at the soil–atmosphere interface and prevents plant growth. Nonetheless, even those covered soils may still be functioning and providing benefits. For example, water from leaky sewage and potable water pipes, or from cracks in deteriorating impervious areas allow water to infiltrate and flow through soil which acts as an *"urban karst"* (landscape on limestone characterized by sinkholes, ravines and underground streams) structure to urban hydrogeology systems (Kaushal and Belt 2012). The ecosystems services provided by the soil are the storage and filtering of the water.

Given the disturbance to urban soils, there is an increased need for engineered soils to manage runoff and provide a lightweight growing medium for vegetation. These are manufactured from a variety of components, such as sand, silt, clay, organic matter, biochar, ash and other soil conditioners to make suitable for a specific application. These soils can be used for green infrastructure such as green roofs, rain gardens or bio-retention basins, or for other purposes such as constructing road beds. Engineered soils are slowly emerging as a solution to problems where the use of natural soils may not be feasible; for example, an engineered soil, Stalite Soil Blend which is a light-weight soil and a medium for plant growth, was used to reduce the weight load on a below-ground parking structure (www.stalite.com/). Another example is PermaTill which is light-weight soil to reduce load on green roofs (www.permatill.com/).

Chemical properties

In addition to physical disturbances, urban air pollution, nutrient and contaminant deposition, chemical spills and applications affect soils, and therefore human and plant health. Sources and concentrations vary with cities and depend on the age and history of industry at a region. Heavy metals, such as Pb (lead), Cu (copper), Ni (nickel), Cr (chromium), Hg (mercury), Zn (zinc), and Cd (cadmium) are the most studied contaminants. Some of these may come from the parent material, and from past industrial activities. The most significant sources of surface soil contamination are automobile exhaust, paint, industry, air pollution, and pesticide use; also, there is contaminated runoff onto soils from impervious areas (Kaushal *et al.* 2017). The extent and spatial variability of metal contamination has been extensively studied; much less is known about the effects of organic contaminants and other less studied pollutants such as asbestos and synthetic chemicals (Bernhardt *et al.* 2017). Organic pollutants also have multiple sources, which include vehicular emission, coal combustion, biomass burning, and waste disposal. Some organic pollutants are persistent, thus soil can be a long-term sink for these compounds. Others are more mobile and can migrate to water bodies causing ecological or health problems there (Yang and Zhang 2015). Organic pollutants are a serious concern in developing countries where the rate of urbanization is currently highest.

Urban soils are usually more alkaline than the corresponding native soil (Pouyat *et al.* 2015b), which is attributed to usage of concrete and cement from construction either as fill material and dust deposition. In areas where natural soil is acidic, residential lawns, golf courses, and other turf areas are routinely amended with lime to promote grass growth. Soil pH also affects metal availability and thus metal uptake by plants and soil invertebrates.

In addition to the physical effects of removing top soil, there are chemical consequences with top soil removal. Nitrogen, phosphorus, and potassium are decreased when organic matter is stripped away for development. Lawn fertilization and irrigation affect the nutrient status of the soil, and, if excessive chemicals are used, there is a danger of possible leaching to surface and groundwater, eventually causing eutrophication (increased productivity, often algal blooms) (Kaye *et al.* 2006). In cold climates road deicers can also cause stress to vegetation and soil fauna.

Urban soil communities

Soil is essential to support life on land. Soil itself is alive and a habitat for a diverse set of organisms that spend part of or their entire life cycle belowground. The abundance of life belowground is astounding; for example, besides hundreds of millions of microbes (e.g. bacteria and fungi) in one square meter of soil, hundreds of earthworms, millipedes, isopods, tens of thousands of springtails (Collembola), hundreds of thousands of mites, and millions of nematodes can coexist. Because all major terrestrial invertebrate groups are represented in the soil, species richness (total number of species) of soil communities is even more amazing. Due to its enormous diversity, the soil ecosystem has been termed the "poor man's rainforest" (Giller 1996). For simplicity, soil fauna are often divided into three major categories based upon size (Swift *et al.* 1979): microfauna (primarily Protozoa, and Nematoda), mesofauna (springtails, mites, larger nematodes, and Enchytraeidae commonly called potworms), and macrofauna (most insects, millipedes, isopods, earthworms, spiders, snails and slugs) (Figure 2.2). Species richness is only one of many dimensions of biodiversity. Biodiversity, measured at other levels of biological organization, is referred to as genetic or ecosystem diversity. Phylogenetic diversity recognizes evolutionary differences among species, while functional diversity focuses on the various roles species play in ecosystem function (Webb *et al.* 2002; Petchey and Gaston 2006). Functionally, soil biota are categorized into ecosystem engineers (mostly earthworms, ants, termites, and vertebrates), biological regulators (mostly nematodes and mesofauna), and chemical transformers (mostly bacteria and fungi) (Turbé *et al.* 2010).

The diversity of life belowground is related to the structural complexity, patchiness of resources, and diverse microclimatic conditions, all of which are changing over time. This heterogeneity exists at many spatial scales from microns to meters, creating favorable conditions to different taxa and allowing for their coexistence. A recent survey in Central Park, New York City (Ramirez *et al.* 2014) illustrated that diverse microbial communities exist in urban settings: molecular analysis of hundreds of soil samples revealed nearly as many phylotypes (environmental DNA sequences showing high similarity) within this single park as are presently known across the globe. This high diversity was partially due to the diverse soil conditions across the park: for instance, variation in soil pH largely explained high microbial community turnover (dissimilarity) among samples. Similarly, Epp Schmidt *et al.* (2017) found that soil bacterial communities in urban areas across five globally

FIGURE 2.2 Urban soils harbor a diversity of organisms. From top to bottom: ground beetles (Insecta, Carabidae), millipedes (Arthropoda, Diplopoda), nematods (Nematoda), ants (Insecta, Hymenoptera), harvestmen (Arachnida, Opiliones), mites (Arachnida, Acari), earthworms (Annelida, Clitellata), springtails (Arthropoda, Collembola), fungi (Fungi, Ascomycota), centipede (Arthropoda, Chilopoda), scorpions (Arachnida, Scorpiones), woodlice (Crustacea, Isopoda). Photos courtesy of Chih-Han Chang (earthworm), Walter P. Pfliegler (nematode, fungus), Zsolt Ujvári (all the others). Used with permission.

distributed cities were not vulnerable to biodiversity loss and that soil pH was the primary soil factor affecting diversity.

Urban soils are often portrayed as "lifeless dirt," which cannot be farther from the truth. In fact, urban and industrial soils may exhibit higher biodiversity and biological activity than assumed based upon their soil physico–chemical characteristics. Joimel *et al*. (2017) compared the topsoil of over 750 sites with five different land uses, and found that arable land and vineyards, not urban land, harbored the least biodiversity of two microarthropod groups, springtails (Collembola) and mites (Acari). This study also highlighted the need for biological assessment of urban soils when evaluating its quality for various uses. Obviously, the diverse patchwork of urban soil systems (Figure 2.1) will harbor a different set of microbial and invertebrate assemblages (Philpott *et al.* 2014; Reese *et al.* 2015; Epp Schmidt *et al.* 2017). At a given locality, community composition is a net result of both natural and anthropogenic factors. We do not know of any major taxon that would be completely excluded from the urban environment as long as the soil is pervious. Certain organisms utilize even the soil under sealed surfaces (pavements, roads, parking lots) as habitat or nesting sites.

Several frameworks have been proposed to highlight the suite of factors creating and maintaining soil biodiversity in urban settings (Pavao-Zuckerman and Byrne 2009; Swan *et al.* 2011). The common thread in these frameworks is that the underlying natural conditions (climate, parent material), site history, past and present disturbances and stress, as well as management practices interact to shape soil community structure and function.

While some habitat patches may harbor a diverse soil community, physical disturbances such as construction involving vegetation and topsoil removal, grading, filling, compaction, and soil sealing represent a major impact for the existing soil biota, and thus lead to local biodiversity loss. Additionally, urban soil conditions can prevent successful colonization and existence of species. As most endogeic (soil dwelling) organisms need to move through the soil, increased compaction and reduced porosity can limit the number of species. Only a small group of macrofauna (ants, earthworms, termites) creates their own passageways. Soil is a semi-aquatic environment, allowing both terrestrial and aquatic organisms to coexist. Soil water content can fluctuate, and both extremely wet and extremely dry soil can be stressful for soil organisms. Soil organic matter, the main resource for true soil dwellers, can be another limiting factor. As outlined above, highly disturbed urban soils are low in SOM, leading to local extinction and slowing down of re-colonization rate by soil biota. Landscaping or soil restoration practices that include amending the soil with organic matter not only benefit plants, but ensure high diversity of belowground biota as well. In terrestrial ecosystems the major source of carbon annually is dead aboveground plant material and dead roots. Removing grass clippings and leaf litter disrupts the natural carbon cycle and deprives the soil fauna of these important resources. At the same time, people create hot-spots by building compost heaps and mulching under planting beds.

Soil pollution is a well-recognized urban issue, which, in addition to being a potential stressor limiting biodiversity, poses serious public health risks. Heavy metal pollution provides a good example of how belowground and aboveground biota is connected. For instance, metals ingested by invertebrates can move through the food chain potentially causing problems in birds and mammals (Reinecke *et al.* 2000; Pouyat *et al.* 2015a).

As discussed above, local physical and chemical characteristics of urban soils are a result of site history, age, past and present disturbances and management. Plants, invertebrates, and microbes are affected by soil conditions, while at the same time they themselves are soil forming factors intimately interacting with the organic and mineral soil components. The net result of these factors is a patchwork of soil conditions and soil community structure at a city scale. The diversity of plant, animal, and microbial communities can be measured at different spatial scales. Local species diversity (expressed as species richness or an index of diversity) is referred to as alpha diversity, while similarity between two or more local communities is called beta diversity, or species turnover. Ultimately all species in an area are a subset of a regional species pool, referred to as gamma diversity. Urbanization is often mentioned as a leading cause of local biodiversity loss (McKinney 2008), that is a smaller subset of the regional species pool can survive in cities than in the more natural habitats, although this general statement appears to be highly taxon and location dependent (Saari *et al.* 2016). Soil invertebrates are small, and thus they can sustain viable populations in a relatively small area, as long as they do not need specific resources or habitats. For these organisms, fragments of native habitats (or remnants), minimally managed parks and gardens, and green corridors, such as stream banks, serve as habitat refuges ensuring their survival in urban landscapes (Vilisics and Hornung 2009).

It is almost impossible to fully assess soil biodiversity at a location, because huge gaps exist in our knowledge of certain taxonomic groups. Species identity is important when there are differences in the role species play in ecosystem function or when the concern is protection of rare or unique species. Soil organisms are not charismatic, and some groups, especially the species rich ones, are often considered to be functionally redundant, meaning that loss of a species from the ecosystem does not affect its function. Whether or not this is the case depends largely on the chosen function and taxa performing that function. For instance large and diverse microbial communities can mineralize carbon, but only specific taxa can fix or mineralize nitrogen. There are also keystone groups or even species, such as earthworms whose presence fundamentally alters virtually all physical, chemical, and biological characteristics of the soil. The relationship between soil biodiversity and function is complex, and experimental data are still scarce. The trend emerging from the studies across different size classes and taxonomical groups is that the assessment of key morphological, physiological, and behavioral features may be more meaningful than simply counting species. This approach, called the "*trait-based approach*," better connects to the many ecosystem functions and services the soil ecosystem provides. The combination of different traits across size classes and feeding groups ensures high functional diversity and promotes ecosystem processes. The trait-based approach does not distinguish between native and introduced species, which, depending on taxonomic group and geographical location, can make up a high proportion of the urban soil fauna. Transporting vast amounts of soil and ornamental or food plants across large distances inevitably results in the introduction of non–native species. Many such species are successful in human environments, and thus have become common in urban settings. They are often referred as synanthropic ('syn' = 'with' 'anthropos' = 'human') species. It is not possible to eliminate them in the highly managed urban settings; however, a better understanding of their ecological roles will help to manage their populations when necessary.

TABLE 2.1 Important functions of the soil ecosystem in urban environments, and the role various groups play in fulfilling those functions. The examples illustrate functions beneficial for humans.

Soil functions	Role of soil biota	Important taxa
Decomposition	Comminution of detritus Chemical breakdown	Isopods, millipedes, ants, earthworms Bacteria and fungi
Nitrogen cycling	N fixation, N mineralization	Specific groups of bacteria
Carbon sequestration	Build up soil organic matter	Bacteria and fungi, plant roots
Infiltration and storage of water	Decompact soil, create macropores, build up soil organic matter	Earthworms, ants, small mammals, plant roots
Medium for plant growth	Improve soil structure Improve nutrient availability Mixing mineral and organic components (bioturbators)	Mycorrhizal fungi, earthworms Fungi and bacteria Earthworms, termites, ants
Resource for wildlife	Food for ground feeding birds, insectivores	Earthworms, larvae, snails
Pest control	Regulating populations of harmful species	Spiders, centipedes, predatory nematodes, mites, beetles
Resistance and resilience	Better withstand disturbances Faster recovery after disturbance	The complex network of soil biota

The Millennium Ecosystem Assessment lists four major ecosystem service categories: provisioning, supporting, regulating, and cultural (MEA 2005). Many soil functions fit in all of those categories in both urban and non–urban settings (Turbé et al. 2010), but in cities some have more importance (Pavao–Zuckerman 2012; Setälä et al. 2014) (Table 2.1). For instance, impervious surfaces increase runoff that large areas of well–drained soil can mitigate. However, compacted soil has low infiltration rates and thus performs poorly during high intensity rain events. Additionally, saturated soil becomes anaerobic (lacks oxygen), which is detrimental for soil organisms and promotes greenhouse gas emissions.

Decomposition, nutrient cycling, and carbon sequestration are the main supporting services the soil ecosystem provides. The basis of the soil food web is plant detritus in the form of leaf litter, thatch, dead roots, and dead wood. The continuous or seasonal input of these resources is essential to maintain soil fertility and a healthy soil community. Soil organisms can on the one hand release nutrients from detritus, and on the other hand incorporate carbon in deeper layers, thereby building up healthier topsoil. For aesthetic or other reasons, dead plant material is often removed from the soil surface, thereby disrupting the natural nutrient cycle. They then replace the nutrients or organic cover by amending fertilizers and mulch. Recycling dead plant material in its place of origin reduces the need for fertilizers. Preserving leaf and woody detritus on the soil surface and landscaping with diverse plant structural types enhance habitat complexity. This results in higher belowground diversity, increased decomposition rates and reduced storm-water runoff (Byrne 2007; Ossola *et al.* 2015; Ossola *et al.* 2016).

Urban agriculture is the most obvious provisioning soil ecosystem service that has been steadily gaining momentum both in developing and developed nations. Over 800 million people practice some form of urban farming or animal husbandry worldwide (FAO 2017). Urban farming increases food security and provides better access to more nutritious food especially for low income residents, while at the same time puts abandoned land to beneficial use. Perhaps less recognized, but potentially important ecosystem services of urban soil are the recreational and educational opportunities it provides for residents. Soil, "*the poor man's tropical rainforest*" is an excellent medium to demonstrate biodiversity, to connect ecosystem and human health, and to discuss sustainability issues. For many inner city residents venturing out to their neighborhood parks or studying the schoolyard is often their only "encounter" with nature.

An ecosystem function unique to cities is food waste removal. In New York City, Youngsteadt *et al.* (2015) found that arthropods, mainly ants, remove large quantities of discarded food thereby making it less available to rats and other less desirable fauna. This ecosystem service of soil invertebrates was especially significant in small green spaces such as medians. Soil invertebrates also constitute a main food source for organisms that people care about, such as birds. For example the rusty blackbird (*Euphagus carolinus*), a species vulnerable to extinction in the IUCN Red List, now utilizes suburban landscapes in the southeastern United States as wintering grounds. There earthworms make up the largest proportions of their winter diet (Newell Wohner *et al.* 2016).

In summary, diversity and abundance of soil communities in urban environments, similarly to more natural settings, reflect the physical chemical characteristics of the local soil, while at the same time actively participating in soil evolution. The soil biota is diverse, and soil organisms are essential to maintain vital ecosystem functions from recycling nutrients and waste to being a resource for other species aboveground.

The urban soil landscape is diverse

The most distinguishing feature of the urban landscape is its spatial heterogeneity, mostly resulting from dividing and subsequently manipulating the land by different owners. The variety of urban soil profiles adds to the already diverse urban landscape (Figure 2.1), and is the main reason for the lack of urban soil classification systems that include the use and management of soils. (Capra *et al.* 2015).

The major characteristics of naturally forming soils are predictable if climate, parent material, age, and biota are known. This is not the case in the urban setting. An area covered with grass may look uniform from the bird's-eye view; however, underneath, the soil profile can greatly vary (Figure 2.1). Soil biota on this fragmented landscape varies even more, making predictions of soil functions, such as infiltration and decomposition rates, challenging. "Digging deeper," i.e. characterizing existing pervious areas in all three dimensions will facilitate understanding and sustainably managing them.

While the patchy urban soil landscape shows large variation in soil physical, chemical, and biological features at city scale (Joimel *et al.* 2016, Ossola and Livesley 2016), soils across cities may exhibit common characteristics as a result of similar human manipulations and management. For instance, urban topsoils often have higher pH values and higher metal content than their rural counterparts (Pouyat *et al.* 2015a, Joimel *et al.* 2016). In other words, urban soils across the globe may be more alike than undisturbed global soils.

Urban soil management

As previously mentioned, soils can provide many ecosystem services in urban areas. Indeed, urban soils have been referred to as the *"brown infrastructure"* of cities and towns (Pouyat *et al.* 2007). Several studies have revealed that the potential for soil to provide ecosystem services in urban areas is surprisingly high, but only if the soil is managed appropriately (e.g. Schwartz and Smith 2016), or if various soil processes can benefit from some urban environmental changes, such as atmospheric deposition of nitrogen, calcium, or other elements. Nonetheless, the services provided by soil in urban areas can have significant limitations, as much of the soil is covered by impervious surfaces or has been recently disturbed or contaminated by human activity.

When an existing area is targeted for remediation, restoration, and reuse, assessment of the existing soil is necessary. This is particularly true for urban areas because the soil forming factors are overwhelmed by human impacts, and thus hard to predict at this stage of our understanding. The effort and complexity of this assessment depend on the planned use and ecosystem services the site is expected to provide. For reducing runoff, deeper soil conditions have to be described, while for planting flowers, assessing the topsoil might be sufficient. Biological surveys are rarely done, yet they provide a more complete picture of the ecological potential, i.e. the possible ecosystem services of a site in addition to the routine soil physico-chemical assessment (Joimel *et al.* 2017).

In the shrinking cities of North America, such as Detroit, Cleveland, and Baltimore, vacant lots occupy increasingly more space of the urban landscape. For instance, in Baltimore over 17,000 vacant lots exist and this number is likely to increase due to the equally high number of vacant properties, many of which face demolition. Vacant lots have unsealed soil, and thus a large potential for achieving the ecosystem services mentioned above (Herrmann *et al.* 2016). Plans for beneficial reuse of these lots include recreational green spaces and urban gardening. The soil on these lots will likely vary due to former site uses, current vegetation cover, characteristics of the existing soil, and other environmental factors. Understanding exposure and human health risks of contaminated soils informs on the potential of vacant lots for redevelopment and can facilitate the crafting of guidance, and possibly even regulation in support of assisting future site users. Selecting the best soil health indicators, whether biological, chemical, or physical that match the planned purpose, is essential. However, risk assessment is often carried out using worst case scenarios, and site level evaluation is not conducted because it requires expertise, time, and money. Moreover, currently there are no universal soil quality based risk standards on contaminant levels. Decision support tools can help planning for reuse and remediating of brownfields. In the course of planning, special attention has to be paid to the perceptions, needs, health, and welfare of the residents who will reuse these abandoned lots (Nassauer and Raskin 2014). A user friendly guide and decision support tool was developed by Toronto Public Health for vacant lot reuse (Archbold and Goldacker 2011). Steps needed to evaluate soil for gardening are illustrated in Figure 2.3.

Intuitively, we understand that healthy soil is the foundation for plant growth and various ecosystem functions, yet there is a general lack of awareness of the role soil biodiversity, from microbes to macrofauna, plays in sustaining these ecosystem functions. Perhaps this lack of awareness, coupled with existing gaps of knowledge, explains why to date no legislation or even regulations exist to protect and enhance soil biodiversity (Turbé 2010). Nonetheless, we know enough of the ecology of these

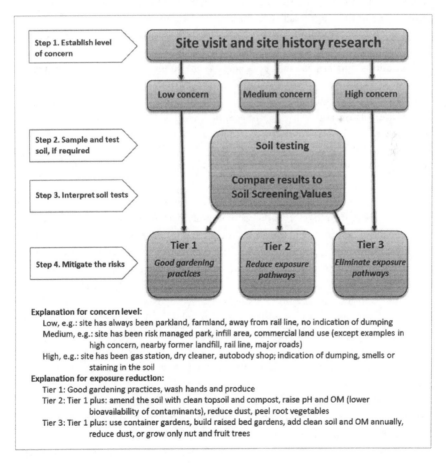

FIGURE 2.3   Decision support tool for assessing urban soil for gardening. Modified from Archbold and Goldacker (2011). Used with the permission of Toronto Public Health.

organisms to understand and value the ecosystem services they provide. Pimentel and colleagues (1997) estimated the total economic benefits of biodiversity with special attention to the services soil biodiversity provides (soil formation, organic matter and nutrient cycling, pest control, etc.) to be one trillion US Dollars annually worldwide.

Traditionally, landscaping is plant-centered with little consideration to the strong linkages between the soil ecosystem and plant growth. Much of the above-ground physical management affects the soil, and this feeds back to the existing vegetation. For instance, during landscaping, grading removes organically rich topsoil, heavy equipment compacts the soil (reducing infiltration and drainage), excavation can damage roots, and the introduction of pavement (even if it has an eco-friendly pervious surface) changes movement of water.

Whether changing the green infrastructure aboveground, restoring degraded soil (brownfields), or creating a novel habitat (green roofs), recognizing these inter-actions is essential if our goal is an urbanized landscape that is environmentally, economically, and socially sustainable in the long term. Byrne and Grewal (2008) termed this holistic approach "ecological landscaping" and provided fundamen-tal questions and guidelines. It explicitly includes social, cultural, and monetary considerations while also emphasizing scientific understanding of the environmen-tal constraints and larger scale ecological consequences in landscape design and restoration. This type of system thinking should take into account a sustainable plant-soil system including the role of soil biodiversity. A concrete example of this is inclusion of soil fauna in designing raingardens. Soil invertebrates can substan-tially alter plant growth, infiltration, and the availability of nutrients and pathogens. Incorporating them in the design and testing their effects potentially improves the efficiency and increases the longevity of raingardens in these natural stormwater treatment systems (Mehrig and Levin 2015).

Managing soil biodiversity is challenging. Unlike plants and other macro-organisms aboveground, soil biodiversity is rarely managed directly. The most obvi-ous example is pest management in lawns, vegetable gardens, and areas surrounding houses. Unfortunately, most pest control companies use chemicals that do not selectively target specific groups of pests, often killing harmless or even beneficial species. For more ecosystem based pest control strategies one has to learn about the local fauna and the role they play in that particular ecosystem to decide whether or not population control is needed. For instance, while earwigs (Insecta: Dermaptera) may cause damage on some fruits and crop plants, they also feed on aphids and other soft bodied insects, thus acting as biological control agents. In general, they do not cause major harm in urban gardens, but they may have to be managed in orchards. Science-based sources, such as the University of California IPM (Integrated Pest Management) website provides reliable information about the most common soil invertebrates, their ecological role, and the various methods to control them should it be necessary (http://ipm.ucanr.edu/PMG/menu.homegarden.html).

In contrast to elimination of pest species, active promotion of soil organisms rarely happens, and usually involves earthworms such as the deliberate introduction of earthworms to Manhattan Island (now New York City) by the Dutch settlers and using earthworms in composting. Other examples are inoculation of soil with mycorrhizae (usually for tree planting), or use of nematodes as biocontrol agents against insects such as mole crickets, weevils, grubs, and cranefly larvae that potentially damage turf grass.

Most human-induced modification of soil community structure happens via soil and mulch transportation and by altering the soil habitat. Amending soil with fertilizers, lime, organic matter, and irrigated water will boost plant growth, and, inadvertently, affects belowground life. Irrigation is a management practice that most fundamentally affects survival and activity of soil organisms. For the majority of soil fauna, soil moisture is a limiting factor; watering lawns and planting beds reduces daily, weekly, or seasonal fluctuations and creates more favorable soil conditions in which they thrive. Irrigation in water-limited regions may allow existence of species that would not exist in the more extreme or fluctuating natural soil environment, eventually shifting urban soil community composition.

In general, increasing habitat complexity aboveground promotes biodiversity belowground (Byrne 2007; Ossola et al. 2015). Designing green spaces with varying cover types that receive an array of management efforts, or encouraging property owners to vary their management within a parcel, such as a mix of planting beds, turfgrass, or minimally managed sections of property, will result in a diversity of soil habitats and thus a diversity of species. Plant selection is also crucial, as plant functional types affect urban soils and their ability to accumulate carbon and nitrogen (Setälä et al. 2016). Additionally, time plays a crucial role in the development of plant-soil interactions and in promoting belowground biodiversity. In urban parks, concentrations of C, N, and organic matter increase with park age, and, in northern latitudes reaches levels similar to those in natural forests (Setälä et al. 2016). Morriën et al. (2017) showed that as a soil ecosystem recovers from long term anthropogenic use, the soil community shifts during soil development and the soil network becomes more complex, leading to a more efficient carbon sink.

It would be beneficial for planners to develop a typology of soil habitat types that are based on management, disturbance, and age such that the greatest variety of habitats can be created in an urban landscape for any period of time. As an example, the Global Urban Soil Ecology and Education Network (GLUSEEN) has developed a typology of six urban soil habitat types based on disturbance and management intensities (Pouyat et al. 2017). In this typology, the soil habitat types approximately correspond to a continuum of human impacts from relatively low influences (native) to those somewhat impacted by urban environmental effects such as remnant forest patches (Pouyat et al. 2009), to types that are highly altered by physical disturbances and management. The latter include highly disturbed soils without structure and low in organic matter content, engineered soils, such as green roof media and street tree pit soils and soils that were once drastically disturbed, but are now managed, such as public or residential lawns. Intermediate in impact are habitat types that are managed but experience relatively low disturbance such as perennial gardens. Landscape and urban designers can use this typology of soil habitat types to optimize the potential for a diversity of soil organisms to occur in an urban area.

Conclusions

In cities there is an inherent diversity of soil conditions from remnants of naturally developed to fully engineered soils. They represent a variety of habitats for soil organisms from marginal to optimal conditions. Plants and the soil ecosystems are intimately linked, and, from a human perspective, soil biota harbors both beneficial and harmful species. Regardless, a diverse soil biota is essential to maintain long-term soil health in the urban landscape. Landscaping and management practices that foster aboveground habitat complexity result in a more diverse, more connected, and better functioning soil biota. Research on managed landscapes provides insight on the development and resilience of soil communities and their importance in maintaining plant health and hydrologic functions. Evaluations of soil conditions prior to reuse are fundamental and require user friendly guides and decision tools. Monitoring soil conditions, including soil biota, on the highly fragmented urban landscape is time- and labor-intensive, providing citizen scientists an increasingly important role in collecting data. Students, residents, and organizations can make observations, and take measurements or photographs. By sharing this information they contribute to a much larger dataset that a small group of researchers could not collect by itself. Active participation may lead to residents appreciating the soil as a valuable natural resource, becoming stewards of their local ecosystem, and employing sustainable management practices.

Acknowledgement

We thank the editors, Alessandro Ossola and Jari Niemelä for inviting us to contribute to this book. We are grateful to Ken Belt for his helpful comments on an earlier version of this manuscript. This material is partially based upon work supported by the National Science Foundation under grants DEB-1027188 and ACI 1244820.

References

Archbold J. and Goldacker S. (2011) *Assessing Urban Impacted Soil for Urban Gardening: Decision Support Tool.* Technical Report and Rationale. Toronto Public Health pp. 114.

Arnold R.W. and Riecken F.F. (1964) Grainy gray ped coatings in Brunizem soils. *Proceedings of the Iowa Academy of Sciences* 71: 350–360.

Bernhardt E.S., Rosi E.J. and Gessner M.O. (2017) Synthetic chemicals as agents of global change. *Frontiers in Ecology and the Environment* 15: 84–90.

Byrne L.B. (2007) Habitat structure: A fundamental concept and framework for urban soil ecology. *Urban Ecosystems* 10: 255–274.

Byrne L.B. and Grewal P. (2008) Introduction to ecological landscaping: A holistic description and framework to guide the study and management of urban landscape parcels. *Cities and the Environment* 1(2): article 3. Available at http://escholarship.bc.edu/cate/vol1/iss2/3 (accessed 1 July 2017).

Capra G.F., Ganga A., Grilli E., Vacca S. and Buondonno A. (2015) A review on anthropogenic soils from a worldwide perspective. *Journal of Soils and Sediments* 15: 1602–1618.

Chen Y., Day S.D., Wick A.F., Strahm B.D., Wiseman P.E. and Daniels W.L. (2013) Changes in soil carbon pools and microbial biomass from urban land development and subsequent post-development soil rehabilitation. *Soil Biology and Biochemistry* 66: 38–44.

Epp Schmidt D.J., Pouyat R., Szlavecz K., Setälä H., Kotze D.J., Yesilonis I., Cilliers S., Hornung E., Dombos M. and Yarwood S.A. (2017) Urbanization leads to the loss of ectomycorrhizal fungal diversity and the convergence of archaeal and fungal soil communities. *Nature Ecology and Evolution* 1: 0123.

FAO (2017) Food and Agriculture Organization of the United Nations. Available at www. fao.org/urban-agriculture/en/ (accessed 11 March 2017).

Giller P.S. (1996) The diversity of soil communities, the 'poor man's tropical rainforest'. *Biodiversity and Conservation* 5: 135–168.

Herrmann D.L., Shuster W.D. and Garmestani A.S. (2015) Vacant urban lot soils and their potential to support ecosystem services. *Plant Soil* 413: 45–57.

ICOMANTH (1997) International Committee on Anthropogenic Soils. Available at www.nrcs.usda.gov/wps/portal/nrcs/detail/soils/survey/class/taxonomy/?cid=stelprdb 1262283 (accessed 11 May 2017).

Jenny H. (1941) *Factors of Soil Formation: A System of Quantitative Pedology.* McGraw-Hill, New York.

Joimel S, Cortet J., Jolivet C.C., Saby N.P.A., Chenot E.D., Branchu P., Consalès J.N., Lefort C., Morel J.L. and Schwartz C., (2016) Physico-chemical characteristics of topsoil for contrasted forest, agricultural, urban and industrial land uses in France. *Science of the Total Environment* 545: 40–47.

Joimel S., Schwartz C., Hedde M., Kiyota S., Krogh P.H., Nahmani J., Pérès G., Vergnes A. and Cortet J. (2017) Urban and industrial land uses have a higher soil biological quality than expected from physicochemical quality. *Science of the Total Environment* 584–585: 614–621.

Kaushal S.S. and Belt K.T. (2012) The urban watershed continuum: Evolving spatial and temporal dimensions. *Urban Ecosystems* 15: 409–435.

Kaushal S.S., Duan S., Doody T.R., Haq S., Smith R,M., Newcomer Johnson T.A., Newcomb K.B., Gorman J., Bowman N., Mayer P.M., Wood K.L., Belt K.T. and Stack W. (2017) Human-accelerated weathering increases salinization, major ions, and alkalinization in fresh water across land use. *Applied Geochemistry* 83: 121–135.

Kaye J.P., Groffman P.M., Grimm N.B., Baker L.A. and Pouyat R.V. (2006) A distinct urban biogeochemistry? *Trends in Ecology & Evolution* 21: 192–199.

Lin H. (2011) Three principles of soil change and pedogenesis in time and space. *Soil Science Society of America Journal* 75: 2049–2070.

McKinney M.L. (2008) Effects of urbanization on species richness: A review of plants and animals. *Urban Ecosystems* 11: 161–176.

MEA (Millennium Ecosystem Assessment) (2005) *Ecosystems and Human Well-Being: Biodiversity Synthesis.* World Resources Institute, Washington, DC.

Mehrig A.S. and Levin L.A. (2015) Potential roles of soil fauna in improving the efficiency of raingardens used and natural stormwater treatment systems. *Journal of Applied Ecology* 52: 1445–1454.

Morel J.L., Chenu C. and Lorenz K. (2015) Ecosystem services provided by soils of urban, industrial, traffic, mining, and military areas (SUITMAs). *Journal of Soils and Sediments* 15: 1659–1666.

Morriën E., Hannula S.E., Snoek L.B., Helmsing N.R., Zweers H., De Hollander M., Soto R.L., Bouffaud M.L., Buée M., Dimmers W. and Duyts H. (2017) Soil networks become more connected and take up more carbon as nature restoration progresses. *Nature Communications* 8: 14349.

Nassauer J.I. and Raskin J. (2014) Urban vacancy and land use legacies: A frontier for urban ecological research, design, and planning. *Landscape and Urban Planning* 125: 245–253.

Newell Wohner P.J., Cooper R.J., Greenberg R.S. and Schweitzer S.H. (2016) Weather affects diet composition of rusty blackbirds wintering in suburban landscapes. *Journal of Wildlife Management* 80: 91–100.

Ossola A. and Livesley S.J. (2016) Drivers of soil heterogeneity in the urban landscape. In Francis R.A., Ossola A. and Livesley S.J. (eds), *Urban Landscape Ecology: Science, Policy and Practice*. Routledge, London and New York.

Ossola A., Hahs A.K. and Livesley S.J. (2015) Habitat complexity influences fine scale hydrological processes and the incidence of stormwater runoff in managed urban ecosystems. *Journal of Environmental Management* 159: 1–10.

Ossola A., Hahs A.K., Nash M.A. and Livesley S.J. (2016) Habitat complexity enhances comminution and decomposition processes in urban ecosystems. *Ecosystems* 19: 927–941.

Pavao-Zuckerman M.A. (2012) *Urbanization, Soils and Ecosystem Services*. In Bargdett R.D., Behan-Pelletier V., Herrick J.E., Jones T.H., Ritz K., Six J., Strong D.R., van der Putten W.H. and Wall D.H. (eds), *Soil Ecology and Ecosystem Services*, pp. 270–281. Oxford University Press, Oxford.

Pavao-Zuckerman M.A. and Byrne L.B. (2009) Scratching the surface and digging deeper: Exploring ecological theories in urban soils. *Urban Ecosystems* 12: 9–20.

Petchey O.L. and Gaston K.J. (2006) Functional diversity: Back to basics and looking forward. *Ecology Letters* 9: 741–758.

Philpott S.M., Cotton J., Bichier P. Friedrich R.L., Moorhead L.C., Uno S. and Valdez M. (2014) Local and landscape drivers of arthropod abundance, richness, and trophic composition in urban habitats. *Urban Ecosystems* 17: 513–532.

Pimentel D., Wilson C., McCullum C., Huang R., Dwen P., Flack J., Tran Q., Saltman T. and Cliff B. (1997) Economic and environmental benefits of biodiversity. *BioScience* 47: 747–757.

Pouyat, R.V., Yesilonis I.D., Russell-Anelli J. and Neerchal N.K. (2007) Soil chemical and physical properties that differentiate urban land-use and cover types. *Soil Science Society of America Journal* 71: 1010–1019.

Pouyat R.V., Carreiro M.M., Groffman P.M., Pavao-Zuckerman M.A. (2009). Investigative approaches to urban biogeochemical cycles: New York metropolitan area and Baltimore as case studies. In McDonnell M., Hahs A. and Breuste J. (eds), *Ecology of Cities and Towns: A Comparative Approach*, pp. 329–351. Cambridge University Press.

Pouyat R.V., Szlavecz K., Yesilonis I., Groffman P. and Schwartz K. (2010) Chemical, physical, and biological characteristics of urban soils. In Aitkenhead-Peterson J. (ed.), *Urban Ecosystem Ecology (Agronomy Monograph 55)*, pp. 119–152. ASA-CSSA-SSSA, Madison, WI.

Pouyat R.V., Szlavecz K., Yesilonis I., Wong C.P., Murawski L., Marra P., Casey R. and Lev S. (2015a) Metal contamination of residential soils and soil fauna at multiple scales in the Baltimore–Washington metropolitan area. *Landscape and Urban Planning* 142: 7–17.

Pouyat R.V., Yesilonis I., Dombos M., Szlavecz K., Setälä H., Cilliers S., Hornung E., Kotze J. and Yarwood S. (2015b) A global comparison of surface soil characteristics across five cities: A test of the urban ecosystem convergence hypothesis. *Soil Science* 180: 136–145.

Pouyat R.V., Setälä H., Szlavecz K., Yesilonis I.D., Cilliers S., Hornung E., Yarwood S., Kotze J.D., Dombos M.D., McGuire M.P. and Whitlow T.H. (2017) Introducing GLUSEEN: a new open access and experimental network in urban soil ecology. *Journal of Urban Ecology* 3(1): jux002.

Ramirez K.S., Leff J.W., Barberan A., Bates S.T., Betley J., Crowther T.W., Kelly E.F., Oldfield E.E., Shaw E.A., Steenbock C., Bradford M.A., Wall D.H. and Fierer N. (2014) Biogeographic patterns in below-ground diversity in New York City's Central Park are similar to those observed globally. *Proceedings of the Royal Society B*: 281.

Reese A.T., Savage A., Youngsteadt E., McGuire K.L., Koling A., Watkins O., Frank S.D. and Dunn R.R. (2015) Urban stress is associated with variation in microbial species composition – but not richness – in Manhattan. *The ISME Journal* 10: 751–760.

Reicosky D.C. (2005) Alternatives to mitigate the greenhouse effect: Emission control by carbon sequestration. In *Simpósio sobre Plantio direto e Meio ambiente; Seqüestro de carbono e qualidade da agua*, pp. 20–28. Anais. Foz do Iguaçu, 18–20 de Maio 2005.

Reinecke A.J., Reinecke S.A., Musibono D.E. and Chapman A. (2000). The transfer of lead (Pb) from earthworms to shrews (Myosorex varius). *Archives of Environmental Contamination and Toxicology* 39: 392–397.

Saari S., Richter S., Higgins M. Oberhofer M., Jennings A. and Faeth S. (2016) Urbanization is not associated with increased abundance or decreased richness of terrestrial animals – dissecting the literature through meta-analysis. *Urban Ecosystems* 19: 1251–1264.

Schwartz S.S. and Smith B. (2016) Restoring hydrologic function in urban landscapes with suburban subsoiling. *Journal of Hydrology* 543, Part B: 770–781.

Setälä H., Bardgett R.D., Birkhofer K., Brady M., Byrne L., De Ruiter P.C., De Vries F.T., Gardi C., Hedlund K., Hemerik L. and Hotes S. (2014) Urban and agricultural soils: Conflicts and trade-offs in the optimization of ecosystem services. *Urban Ecosystems* 17(1): 239–253.

Setälä H.M., Francini G., Allen J.A., Hui N., Jumpponen A. and Kotze D.J. (2016) Vegetation type and age drive changes in soil properties, nitrogen, and carbon sequestration in urban parks under cold climate. *Frontiers in Ecology and Evolution* 4: 93.

SSSA (2017) *Soil Science Society of America Glossary of Soil Science Terms*. Available at https://www.nrcs.usda.gov/wps/portal/nrcs/detail/soils/edu/?cid=nrcs142p2_054280 (accessed 11 May 2017).

Swan C.M., Pickett S.T.A., Szlavecz K., Warren P.S. and Willey K.T. (2011) Biodiversity and community composition in urban ecosystems: Coupled human, spatial and metacommunity processes. In Niemelä J., Breuste J.H., Guntenspergen G., McIntyre N.E., Elmqvist T., James P. (eds), *Urban Ecology: Patterns, Processes, and Applications*, pp 179–186. Oxford University Press.

Swift M.J., Heal O.W. and Anderson J.M. (1979) *Decomposition of Terrestrial Ecosystems*. Studies in Ecology, vol. 5. Blackwell, Oxford.

Turbé A., De Toni A., Benito P., Lavelle Pa., Lavelle Pe., Ruiz N., van der Putten W.H., Labouze E. and Mudgal S. (2010) Soil biodiversity: Functions, threats and tools for policy makers. Bio Intelligence Service, IRD, and NIOO, Report for European Commission (DG Environment).

Vilisics F. and Hornung E. (2009) Urban areas as introduction hot-spots and shelters for native isopod species. *Urban Ecosystems* 12: 333–345.

Webb C.O., Ackerly D.D., McPeek M.A. and Donoghue M.J. (2002) Phylogenies and community ecology. *Annual Review of Ecology, Evolution, and Systematics* 33: 475–505.

Yang J-L. and Zhang G-L. (2015) Formation, characteristics and eco-environmental implications of urban soils: A review. *Soil Science and Plant Nutrition* 61(1): 30–46.

Youngsteadt E., Henderson R.C., Savage A.M., Ernst A. F., Dunn R.R. and Frank S.D. (2015) Habitat and species identity, not diversity, predict the extent of refuse consumption by urban arthropods. *Global Change Biology* 21: 1103–1115.

3

URBAN BIODIVERSITY AND ECOSYSTEM SERVICES

Johannes Langemeyer and Erik Gómez-Baggethun

Introduction

Urbanization happens at the cost of global loss of biodiversity (Hoornweg *et al*. 2016) and is causing profound transformations in global ecosystem functioning (McPherson *et al*. 2016). The United Nations (2014) predicts urban population to rise from approximately 54 per cent today to 66 per cent in 2050, which means that cities will host an additional 2.5 billion inhabitants in the next three decades. With current trends, the surface of urban areas worldwide – currently covering less than 3 per cent of the global terrestrial surface – will double by the end of the next decade, with most of this growth taking place in Africa and Asia (McGranahan and Satterthwaite 2014).

A predominantly urban human society poses some major challenges for global sustainability. Modern life in cities is often perceived as decoupled from biodiversity and healthy ecosystems, because the links between biodiversity and human well-being are increasingly hidden by markets, pipelines and transport networks characterizing the growing complexity of modern industrial societies (see Gómez-Baggethun and de Groot 2010). The global trend of rapidly expanding urban areas (Seto *et al*. 2011) involves that a growing share of the world population is decoupling from direct contact with ecosystems and their dynamics, both physically and cognitively. Instead, the generation of many natural goods and services consumed in cities, including the provision of drinking water, energy and food, air purification, waste disposal and recreation, is often provided from distant locations. This disconnection between people and ecosystems causes a loss of awareness for the human dependency on ecosystems, in what Miller (2005) has referred to as the *extinction of experience* (see Shwartz, Chapter 13).

However, recent global assessments demonstrate that while urban dwellers appear to be losing direct contact with biodiversity and ecosystems, the global human society is not *decoupling* from natural resources but actually becoming increasingly dependent on ecosystems and their *natural capital* (Wiedmann *et al.* 2015). Hence, while urbanization and technological progress have fostered the misconception that modern societies are less dependent on nature, actual demands for ecosystem services keep increasing steadily (Guo *et al.* 2010; Gómez-Baggethun *et al.* 2013). Cities depend on vast areas beyond their boundaries to provide both the resources and ecosystem services they consume and to absorb the waste they produce. For example, urban areas are estimated to account for 60 per cent of the global water use (Grimm *et al.* 2008), with a total water consumption of over 500 litres per capita per day in cities such as Buenos Aires, Argentina, San Diego, USA and Dubai, UAE (WCCD 2016). Cities further account for 60–70 per cent of the world's anthropogenic greenhouse gas emissions (McGranahan and Satterthwaite 2014). Urban inhabitants produce about 6 million tons of waste per day globally, which accounts for two to four times the amount of waste produced by their rural counterparts (Hoornweg and Bhada-Tata 2012).

Hence, besides the intrinsic value of species, at least two crucial arguments stand out for maintaining, restoring and enhancing biodiversity in cities. First, human experiences and interaction with urban ecosystems can create stronger awareness among urban citizens for the need of healthy ecosystems and may lead to actions that reduce the global impacts originated in cities and enhance ecosystem stewardship worldwide. Second, diverse urban ecosystems may provide manifold direct benefits to urban citizens and increase the local supply with ecosystem services, which in turn supports the sustainability and resilience of modern cities. Mounting evidence points to enhancing urban biodiversity and ecosystem services as an important strategy toward United Nations Sustainable Development Goal 11: 'Make cities inclusive, safe, resilient and sustainable' (www.un.org/sustainabledevelopment). However, the relation between biodiversity and ecosystem services remains an open research frontier and due to its eminent complexity the evidence base for correlations between biodiversity and the generation of ecosystem services remains thin (Harrison *et al.* 2014; Duncan *et al.* 2016; Ricketts *et al.* 2016).

This chapter aims at providing an overview of current knowledge and future challenges of understanding urban biodiversity in relation to human well-being. With this objective, the remainder of this chapter is divided into four main sections. First, we describe the importance of biodiversity for sustaining urban ecosystem services. Second, we introduce a classification of the most important ecosystem services provided in urban areas. Third, we focus on a more detailed description of ecosystem services provided by three types of urban green infrastructure: urban gardens, urban forests and urban parks. Finally, we provide a summary of the chapter's main conclusions together with a discussion on the scope and limits of urban ecosystems and biodiversity as nature-based solutions to meet urban sustainability challenges.

Urban biodiversity and its importance for humans

Urban biodiversity has been defined as 'the variety of species richness and abundance of living organisms (including genetic variations) and habitats found in and on the edge of human settlements' (Müller *et al.* 2013, p. 125). Many cities embed areas of high biodiversity, both at the levels of species and ecosystem diversity (CBD 2012). This is partly so because since ancient times cities have often been built in biodiversity rich areas, such as sea- and riverfronts, to facilitate transportation and food supply. Although modern urbanization is not primarily driven by these geographical factors any longer, urbanization still takes place within or close to local biodiversity hotspots (Güneralp and Seto 2013). However, urbanization is not necessarily a threat to local biodiversity. Cities are ecosystems in themselves, and the diverse mosaic of urban biotopes provides multiple niche habitats that favour rare and specialized flora and fauna (Boada and Maneja 2016). For example, higher diversity and abundancy of different bees has been observed for urban areas compared to rural areas (Carre *et al.* 2009; Theodorou *et al.* 2016). Species diversity is further enhanced by the intended and unintended human introduction of non-native species to urban ecosystems. For example, 53 per cent of the plant species in Beijing, China, are alien species (Zhao *et al.* 2010). Non-native species richness generally decreases from the core to urban fringes. The amount of non-native species is also lower in smaller cities, while in larger cities, and specifically in older green spaces within them, the amount of non-native species increases (Müller *et al.* 2013).

The Millennium Ecosystem Service Assessment (MA 2005) promoted the protection of biodiversity for the provision of ecosystem services sustaining human health and well-being. Nevertheless, the relation between urban biodiversity, ecosystem functioning and the provision of ecosystem services still constitutes an open research frontier (Pataki *et al.* 2011), and there still lacks generality whether and to what extent higher biodiversity delivers more ecosystem services (Duncan *et al.* 2016; Ricketts *et al.* 2016). As noted by Haase *et al.* (2014), while most of the research on urban biodiversity has focused on the role of species richness as a measure of diversity (see also Ricketts *et al.* 2016), ecosystem functioning – underlying the generation of ecosystem services – also depends on the identities, densities, biomasses and interactions of populations of species within a community (see also Duncan *et al.* 2016). Robust assessments on urban biodiversity from a broader perspective beyond species diversity (such as proposed by the *City Biodiversity Index* developed under the United Nations Convention on Biological Diversity) are urgently needed to underscore the mounting evidence indicating urban biodiversity as critically important for the endurance and sustained provision of ecosystem services (Gómez-Baggethun *et al.* 2013). Specifically, biodiversity is thought to compensate for fluctuations in the population of individual species and the functions they perform within ecosystems (Loreau *et al.* 2001). This resilience-building property of biodiversity enhances the capacity to sustain ecosystem services dependent on given species and functions in the face of disturbance and change (Green *et al.* 2016). This can be illustrated along the example of food. A monotonous environment apt for the cultivation of only a few crops and animals provides less food security as it will be more vulnerable to disturbances such as heavy rains, droughts, or pest outbreaks affecting the main crops. By diversifying responses to disturbances, diverse crop and animal cultivations increase resilience in the supply of food and other ecosystem services (Elmqvist *et al.* 2003).

Urban ecosystem services in urban planning

Urban ecosystems are strongly transformed and so is the biodiversity it embeds. Due to the still lacking understanding of cities as ecosystems, transformations of urban green, blue and grey infrastructure often happen based on a limited comprehension of urban biodiversity. Only recently, as part of the global push towards more sustainable cities, *urban green infrastructure* strategies have been gaining momentum in policy and planning, especially in the US and Europe (Lennon 2014). The concept of green infrastructure defines 'a strategically planned network of natural and semi-natural areas with other environmental features designed and managed to deliver a wide range of ecosystem services (EC 2013). Hence, urban green infrastructure describes planned urban green areas as well as blue urban subsystems (coastal zones, rivers and standing water) interrelated with their rural counterparts and with the urban grey infrastructure (built-up areas). In contrast to the latter, urban green infrastructure cannot be understood *mechanistically* in supporting single purposes but derives its importance from the multifunctionality and the multiple services sustained by complex socio-ecological interactions. The ecosystem service approach was introduced as a transdisciplinary framework to describe the contribution of ecosystems to human well-being (TEEB 2010). Urban ecosystem services can be defined as the benefits humans derive from urban green infrastructure and other unplanned green and blue spaces in cities.

Bolund and Hunhammar (1999) were the first to develop a systematic description of urban ecosystem services and more comprehensive classifications have been developed over recent years (Gómez–Baggethun and Barton 2013; Gómez–Baggethun *et al.* 2013). Following the MA (2005) and TEEB (2010), urban ecosystem services are generally divided into habitat/sustaining, provisioning, regulating and cultural ecosystem services, a classification that is broadly consistent with the Common International Classification of Ecosystem Services (CICES, http://cices.eu/). Provisioning services describe the production of material matter by urban green spaces to humans, including food, drinking water and raw material. Regulating services enclose the regulation of air quality, local temperature regulation, moderation of flooding events and erosion prevention by urban ecosystems. These services are currently receiving most attention, both by research and by urban planning (Haase *et al.* 2014). Cultural services are the nonmaterial or intangible benefits urban citizens obtain from their interaction with and within urban green spaces, for example, through spiritual enrichment, cognitive development, learning, reflection and place attachment (i.e. Chan *et al.* 2012). Cultural ecosystem services, especially in the form of recreation, aesthetics and social cohesion, are immanently underlying most strategies for preserving and creating green infrastructure in cities. However, their intangible character makes quantitative research and assessment difficult, and to appropriately account for cultural ecosystem services when assessing multifunctional urban green areas still poses an important research frontier. Finally, supporting services (also referred to as habitat services in the TEEB classification and as maintenance services in the CICES classification) are the core ecological processes and functions needed to sustain all other services, including water and nutrient cycles (i.e. MA 2005). The scale at which supporting services can be understood is generally beyond urban boundaries, for example at the watershed level, and underpins the need for a systemic perspective on cities in relation to their surroundings.

FIGURE 3.1 Physical ecosystem disservice: falling branches from an urban street tree in Barcelona, Spain. Courtesy: Gemma Formenti Gonzales.

Despite multiple benefits urban biodiversity may provide, urban green infrastructure has also been related with several negative effects on human well-being – so called *ecosystem disservices* (Fig. 3.1). Research and to a smaller extend urban green space planning tends to depart from a positive perspective on the implementation of green spaces and bares the risk of overlooking or downplaying potential ecosystem disservices and other negative effects green infrastructure strategies may embed. Ecosystem disservices have sometimes been conflated with the absence of ecosystem services. They are better understood as antagonism to ecosystem services – that means urban ecosystem disservices are the negative effects of urban ecosystems on human well-being (Lyytimäki and Sipilä 2009), including both external biophysical impacts on humans and negative psychological experiences inherent to human relations with the biosphere, as well as negative effects on the human society or parts of it. We consequently divide disservices into *physical, psychological* and *societal disservices*. The first include, for example, pollen causing allergies, bites from mosquitos or other animals, the need for foliage removal, as well as the physical destruction of human infrastructure by wild animals, such as wild boars or elephants, or falling tree branches. *Psychological disservices* describe negative feelings, such as fear and disgust related to nature, including animal phobias, as well as immanent feelings of unsafety in urban parks and forests for example at night. Finally, *societal disservices* describe the negative impacts that are indirectly linked to urban green spaces, such as increased crime rates in urban parks, or *ecological gentrification* describing the displacement or exclusion of vulnerable societal groups as a consequence of the implementation of green infrastructure planning (Dooling 2009). Research on urban ecosystem disservices is still in its infancy (von Döhren & Haase 2015), yet a holistic approach to inform urban green infrastructure planning must not ignore negative aspects related to urban biodiversity. Hence, a further categorization and quantification of potential and actual *ecosystem disservices* needs to become a central piece in future research on urban biodiversity in relation to human well-being.

Ecosystem services across different forms of green infrastructure

Type, quantity and quality of ecosystem services provided in urban areas are strongly dependent on the types of green infrastructure underpinning those services. In this section, we describe ecosystem services provided by three types of green infrastructure: vegetable gardens, urban forests and urban parks.

Vegetable gardens

Historically, urban gardens have been a major source of food for urban settlements (e.g., Barthel and Isendahl 2013). The historical development of cities has gone hand in hand with the establishment of horticultural gardens, either within the cities themselves or in their direct surroundings. Only increasing transport capacities through fossil fuels and the development of cooling systems has allowed for larger geographical distances between cities and their horticultural supply sites. Today, in an era of global markets and large-scale transport systems, in many parts of the world urban gardens have lost the crucial economic role they used to play in the past. However, the economic importance of urban gardening differs strongly across geographical regions and socio-economic contexts. In richer cities that are well integrated into global markets and supply chains the role of urban gardening for food security has often become marginal and gardens primarily serve recreational and cultural purposes. In contrast, in poorer cities, especially in the global South, urban gardens remain essential food supply areas. For example, 90 per cent of all vegetables consumed in Dar es Salaam (Tanzania) and 60 per cent of vegetables consumed in Dakar (Senegal) have been reported to originate from urban agriculture (Jacobi *et al.* 2000; Mbaye and Moustier 2000).

History shows that the importance of urban gardens increases in moments of economic crisis – lowering the purchase capacity – and political crisis – reducing excess to global supply networks. Examples are the increased production of food in Swedish cities during World War II (Barthel *et al.* 2010), or in Havana during the political crisis in the 1990s, caused by the end of the Soviet Union as Cuba's most important trading partner (Altieri *et al.* 1999). On a broader time scale, urban gardens can be seen to serve as nodes of biocultural diversity where traditional knowledge on food production is preserved as well as places where a diversity of cultivated plants is maintained (Barthel *et al.* 2010). In addition to food, gardens provide habitat to pollinators, and a range of other plants and animals. Furthermore, urban gardens have also been described for their

TABLE 3.1 Classification of major ecosystem services from urban gardens. Adapted from Langemeyer et al. (2016) based on the broader ecosystem service categories introduced by The Economics of Ecosystem Services and Biodiversity (TEEB, 2010).

Category	Ecosystem service	Examples	Key references
Provisioning services	Food provision	Production of edible fruits and vegetables	Orsini *et al.* (2014); Pourias *et al.* (2015);
	Provision of medicinal and seasoning plants	Production of herbs and spices	Airriess and Clawson (1994)
	Ornamental plants	Growing of flowers and ornamental trees	Dunnet and Quasim (2000)
Regulating services	Maintenance / improvement of soil quality	Nutrient binding and improvement of soil structure through gardening activities	Bøen *et al.* (2013); Li *et al.* 2009
	Erosion prevention and water retention	Soil fixation by plant roots and water retention by leaves	Edmondson *et al.* (2014); Watts and Dexter (1997)

	Local climate and air quality regulation	Absorption of air pollutants and evapotranspiration by plants	Edmondson *et al.* (2014)
	Pollination and seed dispersal	Nutrient source for bees and other pollinating insects	Andersson *et al.* (2007); Jansson and Polasky (2010), Theodorou *et al.* (2016)
Maintenance services	Refuge for plants and animals	(Niche) habitats for different plants and animals	Pourias *et al.* (2015)
	Maintenance of genetic diversity	Reproduction of (rare) cultivated crops	Barthel *et al.* (2010); Barthel *et al.* (2014)
Cultural services	Recreation and relaxation	Space for disconnection from stressful 'urban life'	Hawkins *et al.* (2011); Kaplan and Kaplan (1989); van den Berg and Custers (2011)
	Physical activity	Exercise through gardening activity	Park *et al.* (2011); Dunnet and Qasim (2000)
	Nature experiences	Observation of seasonal changes and growing cycles	Wilson (1984)
	Environmental learning	Experiences with different gardening practices	Beilin and Hunter (2011); Krasny and Tidball (2009)
	Sense of place and place making	Individual and group identification with urban garden spaces	Tidball *et al.* (2014); Okvat and Zautra (2014)
	Social inclusion	Space for experiencing social empowerment	Anguelovski (2013)

importance as a source of cultural ecosystem services, such as recreation and relaxation, environmental learning, the formation of sense of place and especially as spaces favouring social cohesion (Langemeyer *et al.* 2016; Table 3.1) and health benefits (Alaimo *et al.* 2016). In times of abundance, city planners tend to overlook the importance of urban gardens, and many richer cities erase urban gardens from the cityscape. Yet, the call for resilient cities is a strong argument to preserve and foster traditional practices and seed reproduction in urban gardens.

Urban forests

Inhabitants of ancient Egypt transplanted trees to cities in order to obtain benefits such as shading, aesthetic values and supply of fruits. However, Miller *et al.* (2015) noted that trees were rarely abundant in cities and usually concentrated in areas frequented by the 'upper classes', such as parks and ornamental gardens to which 'lower classes' did not have access. It was not before the eighteenth century in Europe that urban trees became important elements of urban planning, including park trees, residential trees, street trees and tree-lined boulevards (ibid.). With cities growing beyond their walls, urbanization became an even more important driver of deforestation. Yet, relics of forests have also been preserved from urban expansions. For example, in 1742 the Tiergarten in Berlin, which previously served as royal hunting grounds, was transformed by the Prussian king Frederick II into a forested urban park. The incorporation of urban forests and trees into European city planning in the renaissance influenced much of the design of cities around the world. Other examples of important urban forests as relics of the pre-urban landscapes are the Table Mountain National Park in Cape Town, South Africa, the Tijuca Forest in Rio de Janeiro, Brazil and the Singapore Botanic Gardens – all three global biodiversity hotspots.

The services for which urban trees are mostly planted have not changed much since ancient times (Table 3.2). Although the importance of local food supply has decreased in most cities, fruit trees are still very common as residential trees. Public, street and park trees fulfil primarily aesthetic and recreational purposes, and improve the micro-climate through shading and evapotranspiration; both are critically dependent on the type of vegetation (Bowler et al. 2010). Other ecosystem services related to urban trees are noise reduction and runoff mitigation (Escobedo et al. 2011). Especially larger populations of trees, often found in the urban fringes, show some potential to contribute to a reduction of air pollution and to store carbon and thereby mitigate global climate change (Paoletti 2009). Nevertheless the amount of carbon storage is relative small and does not go far in compensating the carbon emissions produced in cities; the highest capacity for absorption of air bound pollutants in cities, including PM, SOx, NO_x and CO (Nowak et al. 2006), has been shown for particulate matter (PM_{10}), a major cause of respiratory illnesses (Baró et al. 2014). Most regulating services, including micro-climate regulation, noise reduction, runoff mitigation and air pollution reduction (except carbon storage), are thereby determined by the leaf area. Evergreen trees are thus most beneficial in providing these kinds of ecosystem services.

Urban forests are also related with important disservices: A psychological disservice is the feeling of fear, especially related to thick and dark forests. Physical ecosystem disservices from urban forests are related to damage to infrastructure caused by the root system of trees and falling branches – the latter also causing severe physical injuries to humans (Fig. 3.1). Trees also emit pollen, causing human allergic reactions. Biogenic volatile organic compounds (VOCs) emitted by trees react in the presence of nitrogen oxides (NO_x) and sunlight to ground-level ozone – and are thus an important factor in producing photochemical smog. However, Tsunetsugu et al. (2010) assume that VOCs from trees may also provide some benefits for human health.

TABLE 3.2 Classification of major ecosystem services related to urban forests and trees. Building upon Baró *et al.* (2014) and Wang *et al.* (2016) using ecosystem service categories as introduced by The Economics of Ecosystem Services and Biodiversity (TEEB, 2010).

Category	Ecosystem service	Examples	Key references
Provisioning services	Food provision	Fruits and nuts from residential trees	Lafontaine-Messier *et al.* (2016)
	Timber provision	Provision of building and fuel	Baines (2000)
Regulating services	Runoff mitigation	Water retention especially by leaves	Escobedo *et al.* (2011)
	Noise reduction	Absorption of sound waves by vegetation Barriers	Ozer *et al.* (2008)
	Global climate regulation	Carbon sequestration and fixation by the biomass of urban shrubs and trees	Kordowski and Kuttler (2010); Davies *et al.* (2011); Paoletti *et al.* (2011)
	Air quality regulation	Removal and fixation of air pollutants (e.g. NO_x, SO_4, PM_{10}) by tree leaves	Nowak *et al.* (2006); Baró *et al.* (2014)
	Local climate regulation	Urban cooling through shading and evapotranspiration	Yuan and Bauer (2007)
Maintenance services	Maintenance of biodiversity	Habitat for species such as birds, bats and insects	Kong *et al.* (2010)
Cultural services	Amenity	Aesthetic appreciation of urban trees	Tyrväinen and Miettinen (2000); Yang and Webster (2016)
	Recreation	Stress reduction in urban forests	Hammitt (2000); Arnberger (2006)

Urban parks

A third highly important type of urban green infrastructure supporting human well-being in cities is urban parks (Table 3.3). Konijnendijk *et al.* (2013) define urban parks as 'delineated open space areas, mostly dominated by vegetation and water, and generally reserved for public use'. Urban parks include large emblematic parks such as the Fuxing Park in Shanghai, Central Park in New York City, Tempelhofer Feld in Berlin, Parque Nacional in Bogotá and the Royal Botanic Garden in Melbourne. Yet, most urban parks are of smaller size, integrated into neighbourhoods and rather attracting local visitors. Beyond their size, urban parks may largely differ in their facilities, green space structure and management. Konijnendijk *et al.* (2013) provide a comprehensive review of the scientific evidence base describing benefits for human well-being in relation to services from urban parks.

TABLE 3.3 Classification of major ecosystem services related to urban parks. Building upon Konijnendijk *et al.* (2013) using ecosystem service categories as introduced by The Economics of Ecosystem Services and Biodiversity (TEEB, 2010).

Category	Ecosystem service	Examples	Key references
Provisioning services	–		–
Regulating services	Stormwater and runoff regulation	Buffer zones for flooding events and water retention by vegetation	Kubal *et al.* (2009)
	Local climate regulation	Buffer of extreme heat events through plant evapotranspiration	Bowler *et al.* (2010)
Maintenance services	Promotion of biodiversity	Local habitats for urban flora and fauna including mammals	Cornelis and Hermy (2004)
Cultural services	Recreation and leisure	Space for diverse recreational and leisure activities (e.g. jogging, playing)	Maas (2006); Hussain *et al.* (2010); Stodolska *et al.* (2011)
	Nature experiences and environmental learning	Observation and contact with urban flora and fauna	Langemeyer *et al.* (2015)
	Social cohesion	Space for social interaction	Fan *et al.* (2011); Peters *et al.* (2010)

The most important ecosystem services affiliated with urban parks are recreation and leisure activities, which promote various health benefits, including psychological well-being and mental health, reduced stress, concentration capacity, reduced obesity and general perceptions of quality of life (Maas *et al.* 2006). Urban parks also allow for experiences of nature and biodiversity. In this context, the variety of parks, their management and naturalness is important. A study by Langemeyer *et al.* (2015) indicates that medium management intensities create the best environment for nature experience and environmental learning, while highly managed park areas and unmanaged areas are less likely to provide this service. Urban parks are also described as spaces that provide opportunities for social cohesion and the creation of social identity, in the sense that parks are spaces where people can meet and establish social relationships, yet a scientific evidence base that this potential is actually realized is still missing (Konijnendijk *et al.* 2013). Although urban parks' major role is recreational, they have been described as an important refuge for biodiversity in urban environments (Cornelis and Hermy 2004). Konijnendijk *et al.* (2013) conclude that, especially due to their fragmentation and embedment of diverse microhabitats, most urban parks show increased species richness compared to other green spaces in cities. Enhanced levels of species diversity are further favoured in larger parks that are well integrated into a network of other green spaces. Urban parks often host considerable forest patches with the respective benefits especially from regulating services as described above. Other important regulating services provided by urban parks are groundwater recharge and reduction of surface runoff, which buffers extreme rain events and protects from flooding. However, little research has focused on groundwater recharge and reduction of surface runoff by urban parks (Ossola et al. 2015). To the contrary, urban cooling is much in the focus of urban policies and planners. During daytime average temperatures in urban parks can be up to 1°C lower than in surrounding areas. Besides evapotranspiration and shading by plants, the size of the parks has been found to be crucial for reduced temperatures in urban parks. In summary, it can be stated that creation of urban parks may especially stipulate the provision of cultural services for urban dwellers and some crucial regulating services. Parks are generally perceived as positive green infrastructure elements that enrich the living environment of cities, and an increasing number of studies are showing that the cultural attractiveness of urban parks also stipulates tourism and housing prices and rents (e.g. Brander and Koetse 2011). Increased tourism and housing prices might benefit local economies, yet, they may also favour gentrification processes with potential negative social consequences for the local population. Research on *green gentrification* by urban parks is another research frontier that deserves further attention in the discussion about inclusive urban ecosystems. Studies especially from American and Asian cities do also report on other drawbacks related to urban parks. In the US, specific parks have been related with feelings of insecurity and enhanced crime levels, while negative effects from parks in China and Japan have rather been attributed to noise and light emissions (Konijnendijk et al. 2013).

Conclusions

This chapter gives an overview of the relation between urban biodiversity, ecosystem services and human well-being in cities.

Within this chapter we highlight that cities do not necessarily stand in contrast to maintenance and enhancement of biodiversity, either in terms of species diversity and abundance or in terms of biotope diversity. Yet, the dichotomous understanding of cities on the one hand side, and nature on the other, has for too long blinded the important potential cities have within a global sustainability agenda. A new understanding of cities as coupled social–ecological systems (Niemelä et al. 2011) and integrated parts of the planet's ecosystem (Langemeyer 2015) may help in activating this potential – and must thus be imminent to an emerging *Science of Cities* (McPhearson et al. 2016).

Moral values and the *reconnection to nature* to enhance environmental stewardship are important arguments for the protection of urban biodiversity but too often fall short in providing practical results in policy and planning. Beyond moral arguments, an ecosystem service approach to urban biodiversity underpins its critical importance for human health and well-being in cities. Although wide evidence for the benefits humans obtain from urban biodiversity is readily available, most of the knowledge is still fragmented. For a rigorous integration of biodiversity and ecosystem services into urban green infrastructure planning, more holistic assessments are needed that highlight the multifunctional character and the multiple (co-)benefits related to the creation and protection of urban green areas. As shown in this chapter, the types of ecosystem services urban green areas support range from the supply of food, reducing air pollution or heat extremes to the provision of space for relaxation and social cohesion. "For a rigorous integration of biodiversity and ecosystem services into urban green infrastructure planning, more holistic assessments are needed that highlight the multifunctional character and the multiple (co-) benefits related to the creation and protection of urban green areas."

We have shown examples of how specific types of urban green infrastructure sustain different bundles of ecosystem services. For example, urban gardens are relevant for cultural aspects, such as social cohesion and place-making as well as for food resilience in moments of crisis. Urban forests are most important for regulating services, especially air pollution reduction and urban cooling. Yet, in ancient times they also played important roles in the provision of fuel and timber. Finally, urban parks provide a broad set of cultural ecosystem services, including recreation and social cohesion. Awareness of these specific services can help urban planning to steer benefits to the urban dwellers through green infrastructure strategies.

Today, the idea of urban biodiversity as a nature-based solution is gaining momentum in an attempt to address sustainability challenges in cities. In this context, rigorous integration of urban biodiversity into urban planning through novel green infrastructure strategies not only requires the consideration of multifunctionality and (co-) benefits but also a critical understanding and projection of the needs of the urban society and its different sub-groups now and in the future. This requires considering for whom the benefits are provided, who is potentially excluded from those benefits and who might be negatively affected by physical, psychological and societal ecosystem disservices related to urban biodiversity.

References

Airriess C.A. and Clawson D.L. (1994) Vietnamese market gardens in New Orleans. *Geographical Review* 84(1): 16.

Alaimo K., Beavers A.W., Crawford C., Snyder E.H. and Litt J.S. (2016) Amplifying health through community gardens: A framework for advancing multicomponent, behaviorally based neighborhood interventions. *Current Environmental Health Reports* 3(3): 302–312.

Altieri M.A., Companioni N., Cañizares K., Murphy C., Rosset P., Bourque M. and Nicholls C.I. (1999) The greening of the 'barrios': Urban agriculture for food security in Cuba. *Agriculture and Human Values* 16(2): 131–140.

Andersson E., Barthel S. and Ahrné K. (2007) Measuring social–ecological dynamics behind the generation of ecosystem services. *Ecological Applications* 17(5): 1267–1278.

Anguelovski I. (2013) Beyond a livable and green neighborhood: Asserting control, sovereignty and transgression in the Casc Antic of Barcelona. *International Journal of Urban and Regional Research* 37(3): 1012–1034.

Arnberger A. (2006) Recreation use of urban forests: An inter-area comparison. *Urban Forestry & Urban Greening* 4(3): 135–144.

Baines C. (2000) A forest of other issues. *Landscape Design* 294: 46–47.

Baró F., Chaparro L., Gómez-Baggethun E., Langemeyer J., Nowak D.J. and Terradas J. (2014) Assessing ecosystem services provided by urban forests in relation to air quality and climate change mitigation policies in Barcelona, Spain. *Ambio* 43: 466–479.

Barthel S., Folke C. and Colding J. (2010) Social–ecological memory in urban gardens – Retaining the capacity for management of ecosystem services. *Global Environmental Change* 20(2): 255–265.

Barthel S. and Isendahl C. (2013) Urban gardens, agriculture, and water management: Sources of resilience for long-term food security in cities. *Ecological Economics* 86: 224–234.

Barthel S., Parker J., Folke C. and Colding J. (2014) Urban gardens: Pockets of social-ecological memory. In *Greening in the Red Zone*, pp. 145–158. Springer, The Netherlands.

Beilin R. and Hunter A. (2011) Co-constructing the sustainable city: how indicators help us 'grow' more than just food in community gardens. *Local Environment* 16: 523–538.

Boada M. and Maneja R. (2016) Cities are ecosystems. OurPlanet, the magazine of the United Nations Environment Programme (UNEP).

Bøen A., Haraldsen T. K. and Krogstad T. (2013) Large differences in soil phosphorus solubility after the application of compost and biosolids at high rates. *Acta Agriculturae Scandinavica, Section B-Soil & Plant Science* 63(6): 473–482.

Bolund P. and Hunhammar S. (1999) Ecosystem services in urban areas. *Ecological Economics* 29: 293–301.

Bowler D.E., Buyung-Ali L., Knight T.M. and Pullin A.S. (2010) Urban greening to cool towns and cities: A systematic review of the empirical evidence. *Landscape and Urban Planning* 97(3): 147–155.

Brander L.M. and Koetse M.J. (2011) The value of urban open space: Meta-analyses of contingent valuation and hedonic pricing results. *Journal of Environmental Management* 92(10): 2763–2773.

Carre G., Roche P., Chifflet R., Morison N., Bommarco R., Harrison-Cripps J., Krewenka K., Potts S.G., Roberts S.P.M., Rodet G., Settele J., Steffan-Dewenter I., Szentgyörgyi H., Tscheulin T., Westphal C., Woyciechowski M. and Vaissière B.E. (2009) Landscape context and habitat type as drivers of bee diversity in European annual crops. *Agriculture, Ecosystems & Environment* 133(1): 40–47.

CBD – Secretariat of the Convention on Biological Diversity (2012) *Cities and Biodiversity Outlook*. 64 pages. Montreal, Canada.

Chan K.M.A., Satterfield T. and Goldstein J. (2012) Rethinking ecosystem services to better address and navigate cultural values. *Ecological Economics* 74: 8–18.

Cornelis J. and Hermy M. (2004) Biodiversity relationships in urban and suburban parks in Flanders. *Landscape and Urban Planning* 69(4): 385–401.

Davies Z.G., Edmondson J.L., Heinemeyer A., Leake J.R. and Gaston K.J. (2011) Mapping an urban ecosystem service: Quantifying above-ground carbon storage at a city-wide scale. *Journal of Applied Ecology* 48(5): 1125–1134.

Dooling S. (2009) Ecological gentrification: A research agenda exploring justice in the city. *International Journal of Urban and Regional Research* 33(3): 621–639.

Duncan C., Thompson J.R. and Pettorelli N. (2016) The quest for a mechanistic understanding of biodiversity–ecosystem services relationships. *Proceedings of the Royal Society Series B* 282(1817): 20151348.

Dunnett N. and Qasim M. (2000) Perceived benefits to human well-being of urban gardens. *HortTechnology* 10 (1): 40–45.

Edmondson J.L., Davies Z.G., McCormack S.A., Gaston K.J. and Leake J.R. (2014) Land-cover effects on soil organic carbon stocks in a European city. *Science of the Total Environment* 472: 444–453.

Elmqvist T., Folke C., Nyström M., Peterson G., Bengtsson J., Walker B. and Norberg J. (2003) Response diversity, ecosystem change, and resilience. *Frontiers in Ecology and the Environment* 1(9): 488–494.

Escobedo F.J., Kroeger T. and Wagner J.E. (2011) Urban forests and pollution mitigation: Analyzing ecosystem services and disservices. *Environmental Pollution* 159: 2078–2087.

European Commission (2013) Green infrastructure (GI) – enhancing Europe's Natural Capital. *COM* (2013) 249, Brussels.

Fan Y., Das K.V. and Chen Q. (2011) Neighborhood green, social support, physical activity, and stress: Assessing the cumulative impact. *Health & Place* 17(6): 1202–1211.

Gómez-Baggethun E. and De Groot R. (2010) Natural capital and ecosystem services: The ecological foundation of human society. In *Ecosystem Services* 30: 105–121, The Royal Society of Chemistry.

Gómez-Baggethun E. and Barton D. N. (2013) Classifying and valuing ecosystem services for urban planning. *Ecological Economics* 86: 235–245.

Gómez-Baggethun E., Gren Å., Barton D.N., Langemeyer J., McPhearson T., O'Farrell P., Andersson E., Hamstead Z., and Kremer P. (2013) Urban Ecosystem Services. In Elmqvist T. (ed.). *Urbanization, Biodiversity and Ecosystem Services*, pp. 175–251. Springer, Netherlands.

Green T.L., Kronenberg J., Andersson E., Elmqvist T. and Gómez-Baggethun E. (2016) Insurance value of green infrastructure in and around cities. *Ecosystems* 19: 1051.

Grimm N.B., Faeth S.H., Golubiewski N.E., Redman C.L., Wu J., Bai X. and Briggs J.M. (2008) Global change and the ecology of cities. *Science* 319(5864): 756–760.

Güneralp B. and Seto K.C. (2013) Futures of global urban expansion: Uncertainties and implications for biodiversity conservation. *Environmental Research Letters* 8(1): 014025.

Guo Z., Zhang L. and Li Y. (2010) Increased dependence of humans on ecosystem services and biodiversity. *PLoS One* 5: 1–7.

Haase D., Larondelle N., McPhearson T., Andersson E., Artmann M, Borgström S., Breuste J., Gomez-Baggethun E., Gren Å., Hamstead Z., Hansen R., Kabisch N., Kremer P., Langemeyer J., Lorance E., McPhearson T., Rall E., Pauleit S., Qureshi N., Schwarz N., Voigt A., Wurster D. and Elmqvist T. (2014) Quantitative review of urban ecosystem services assessment: Concepts, models and implementation. *Ambio* 43: 413–433.

Hammitt W.E. (2000) The relation between being away and privacy in urban forest recreation environments. *Environment and Behavior* 32(4): 521–540.

Harrison P.A., Berry P.M., Simpson G., Haslett J.R., Blicharska M., Bucur M., Dunford R., Egoh B., Garcia-Llorente M., Geamănă N., Geertsema W., Lommeleni E., Meiresonnei L., Turkelboom F. (2014) Linkages between biodiversity attributes and ecosystem services: a systematic review. *Ecosystem Services* 9: 191–203.

Hawkins J.L., Thirlaway K.J., Backx K. and Clayton D.A. (2011) Allotment gardening and other leisure activities for stress reduction and healthy ageing. *HortTechnology* 21 (5): 577–585.

Hoornweg D. and Bhada-Tata P. (2012) What a Waste: A Global Review of Solid Waste Management. The World Bank.

Hoornweg D., Hosseini M., Kennedy C. and Behdadi A. (2016) An urban approach to planetary boundaries. *Ambio* 1–14.

Hussain G., Nadeem M., Younis A., Riaz A., Khan M.A. and Naveed S. (2010) Impact of public parks on human life: A case study. *Pakistan Journal of Agricultural Science* 47(3): 225–230.

Jacobi P., Amend J. and Kiango S. (2000) Urban agriculture in Dar es Salaam: Providing an indispensable part of the diet. In Bakker N., Dubbeling M., Guendel S., Sabel Koschella U., de Zeeuw H. (eds), *Growing Cities, Growing Food: Urban Agriculture on the Policy Agenda*, 257–283. DSE, Germany.

Jansson A. and Polasky S. (2010) Quantifying biodiversity for building resilience for food security in urban landscapes: Getting down to business. *Ecology and Society* 15(3): 20.

Kaplan S. and Kaplan R. (1989) *The Experience of Nature: A Psychological Perspective.* Cambridge University Press, New York.

Kong F., Yin H., Nakagoshi N. and Zong Y. (2010) Urban green space network development for biodiversity conservation: Identification based on graph theory and gravity modeling. *Landscape and Urban Planning* 95(1): 16–27.

Konijnendijk C.C., Annerstedt M., Nielsen A.B. and Maruthaveeran S. (2013) Benefits of urban parks: A systematic review. A report for IPFRA. Copenhagen & Alnarp.

Kordowski K. and Kuttler W. (2010) Carbon dioxide fluxes over an urban park area. *Atmospheric Environment* 44(23): 2722–2730.

Krasny M.E. and Tidball K.G. (2009) Applying a resilience systems framework to urban environmental education. *Environmental Education Research* 15(4): 465–482.

Kubal C., Haase D., Meyer V. and Scheuer S. (2009) Integrated urban flood risk assessment–adapting a multicriteria approach to a city. *Natural Hazards and Earth System Sciences* 9(6): 1881–1895.

Lafontaine-Messier M., Gélinas N. and Olivier A. (2016) Profitability of food trees planted in urban public green areas. *Urban Forestry & Urban Greening* 16: 197–207.

Langemeyer J., Baró F., Roebeling P. and Gómez-Baggethun E. (2015). Contrasting values of cultural ecosystem services in urban areas: The case of park Montjuïc in Barcelona. *Ecosystem Services* 12: 178–186.

Langemeyer J., Latkowska M.J., Gómez-Baggethun E., Voigt A., Calvet-Mir L., Pourias J., Camps-Calvet M., Orsini F., Breuste J., Artmann M., Jokinen A., Béchet B., Brito da Luz, P., Hursthouse A., Stępień M.P. and Baležentiene L. (2016) Ecosystem services from urban gardens. In Bell S., Fox-Kämper R., Keshavarz N., Benson M., Caputo S., Noori S. and Voigt A. (eds), *Urban Allotment Gardens in Europe*, pp. 115–141. Routledge; London.

Langemeyer J. (2015) *Urban Ecosystem Services: The Value of Green Spaces in Cities.* PhD-Thesis, Universitat Autònoma de Barcelona & Stockholm University. ISBN 978-91-7649-290-1.

Lennon M. (2014) Green infrastructure and planning policy: a critical assessment. *Local Environment: The International Journal of Justice and Sustainability* 20(8): 957–980.

Li H., Shi W.Y., Shao H.B. and Shao M.A. (2009) The remediation of the lead-polluted garden soil by natural zeolite. *Journal of Hazardous Materials* 169(1): 1106–1111.

Loreau M., Naeem S., Inchausti P., Bengtsson J., Grime J.P., Hector A., Hooper D.U., Huston M.A., Raffaelli D., Schmid B., Tilman D., Wardle D.A. (2001) Biodiversity and ecosystem functioning: Current knowledge and future challenges. *Science* 294(5543): 804–808.

Lyytimäki J. and Sipilä M. (2009). Hopping on one leg–The challenge of ecosystem disservices for urban green management. *Urban Forestry & Urban Greening* 8(4): 309–315.

MA – Millennium Ecosystem Assessment (2005) *Ecosystems and Human Well-being.* Island Press, Washington, DC.

Maas J., Verheij R.A., Groenewegen P.P., De Vries S. and Spreeuwenberg P. (2006) Green space, urbanity, and health: How strong is the relation? *Journal of Epidemiology and Community Health* 60(7): 587–592.

Mbaye A. and Moustier P. (2000) Market-oriented urban agricultural production in Dakar. In: Bakker N., Dubbeling M., Guendel S., Sabel Koschella U., de Zeeuw H. (eds), *Growing Cities, Growing Food: Urban Agriculture on the Policy Agenda*, pp. 257–283. DSE, Germany.

McGranahan G. and Satterthwaite D. (2014) Urbanisation concepts and trends. IIED Working Paper. IIED, London.

McPhearson T., Parnell S., Simon D., Gaffney O., Elmqvist T., Bai X., Roberts D. and Revi A. (2016) Scientists must have a say in the future of cities. *Nature* 538: 165–166.

Miller J.R. (2005) Biodiversity conservation and the extinction of experience. *Trends in Ecology & Evolution* 20(8): 430–434.

Miller R.W., Hauer R.J. and Werner L.P. (2015) *Urban Forestry: Planning and Managing Urban Greenspaces.* Waveland Press, Long Grove, IL.

Müller N., Ignatieva M., Nilon C.H., Werner P. and Zipperer W.C. (2013) Patterns and trends in urban biodiversity and landscape design. In Elmqvist T. (ed.), *Urbanization, Biodiversity and Ecosystem Services: Challenges and Opportunities,* pp. 123–174. Springer, The Netherlands.

Niemelä J., Breuste J.H., Guntenspergen G., McIntyre N.E., Elmqvist T. and James P. (eds) (2011) *Urban Ecology: Patterns, Processes, and Applications.* Oxford University Press, Oxford.

Nowak D.J., Crane D.E. and Stevens J.C. (2006) Air pollution removal by urban trees and shrubs in the United States. *Urban Forestry & Urban Greening* 4: 115–123.

Okvat H.A. and Zautra A.J. (2014) Sowing seeds of resilience: Community gardening in a post-disaster context. In *Greening in the Red Zone,* pp 73–90. Springer, The Netherlands.

Orsini F., Gasperi D., Marchetti L., Piovene C., Draghetti S., Ramazzotti S., Bazzocchi G. and Gianquinto G. (2014) Exploring the production capacity of rooftop gardens (RTGs) in urban agriculture: The potential impact on food and nutrition security, biodiversity and other ecosystem services in the city of Bologna. *Food Security* 6(6): 781–792.

Ossola A., Hahs A.K. and Livesley S.J. (2015) Habitat complexity influences fine scale hydrological processes and the incidence of stormwater runoff in managed urban ecosystems. *Journal of Environmental Management* 159: 1–10.

Ozer S., Irmak M.A. and Yilmaz H. (2008) Determination of roadside noise reduction effectiveness of Pinus sylvestris L. and Populus nigra L. in Erzurum, Turkey. *Environmental Monitoring and Assessment* 144(1–3): 191–197.

Paoletti E. (2009) Ozone and urban forests in Italy. *Environmental Pollution* 157: 1506–1512.

Paoletti E., Bardelli T., Giovannini G. and Pecchioli L. (2011) Air quality impact of an urban park over time. *Procedia Environmental Sciences* 4: 10–16.

Park S.A., Lee K.S. and Son K.C.H. (2011) Determining exercise intensities of gardening tasks as a physical activity using metabolic equivalents in older adults. *HortScience* 46(2): 1706–1710.

Pataki D.E., Carreiro M.M., Cherrier J., Grulke N.E., Jennings V., Pincetl S., Pouyat R.V., Whitlow T.H., Zipperer W.C. (2011) Coupling biogeochemical cycles in urban environments: ecosystem services, green solutions, and misconceptions. *Frontiers in Ecology and the Environment* 9(1): 27–36.

Peters K., Elands B. and Buijs A. (2010) Social interactions in urban parks: Stimulating social cohesion? *Urban Forestry & Urban Greening* 9(2): 93–100.

Pourias J., Duchemin, E. and Aubry C. (2015) Products from urban collective gardens: food for thought or for consumption? Insights from Paris and Montreal. *Journal of Agriculture, Food Systems and Community Development* 5(2): 1–25.

Ricketts T.H., Watson K.B., Koh I., Ellis A.M., Nicholson C.C., Posner S., Richardson L.L., Sonter L.J. (2016) Disaggregating the evidence linking biodiversity and ecosystem services. *Nature Communications* 7: 13106.

Seto K.C., Fragkias M., Güneralp B. and Reilly M.K. (2011) A meta-analysis of global urban land expansion. *PLoS ONE* 6(8): e23777.

Stodolska M., Shinew K.J., Acevedo J.C. and Izenstark D. (2011) Perceptions of urban parks as havens and contested terrains by Mexican-Americans in Chicago neighborhoods. *Leisure Sciences* 33(2): 103–126.

TEEB (2010) The Economics of Ecosystems & Biodiversity: Mainstreaming the Economics of Nature. UNEP.

Theodorou P., Radzevičiūtė R., Settele J., Schweiger O., Murray T.E. and Paxton R.J. (2016) Pollination services enhanced with urbanization despite increasing pollinator parasitism. *Proceedings of the Royal Society B* 283(1833). Available at http://rspb.royalsocietypublishing. org/content/royprsb/283/1833/20160561.full.pdf (accessed 22 August 2017).

Tidball K.G., Weinstein E.D. and Krasny M.E. (2014) Synthesis and conclusion: Applying greening in red zones. In *Greening in the Red Zone*, pp. 451–486. Springer, The Netherlands.

Tsunetsugu Y., Park B.J. and Miyazaki Y. (2010) Trends in research related to 'Shinrin-yoku' (taking in the forest atmosphere or forest bathing) in Japan. *Environmental Health and Preventive Medicine* 15(1): 27–37.

Tyrväinen L. and Miettinen A. (2000) Property prices and urban forest amenities. *Journal of Environmental Economics and Management* 39(2): 205–223.

United Nations (2014) *World Urbanization Prospects: The 2014 Revision, Highlights*. United Nations, Department of Economic and Social Affairs, Population Division. (ST/ESA/ SER.A/352).

van den Berg A.E. and Custers M.H.G. (2011) Gardening promotes neuroendocrine and affective restoration from stress. *Journal of Health Psychology* 16: 3–11.

von Döhren P. and Haase D. (2015). Ecosystem disservices research: A review of the state of the art with a focus on cities. *Ecological Indicators*, 52, 490–497.

Wang H.F., Qureshi S., Qureshi B.A., Qiu J.X., Friedman C.R., Breuste J. and Wang X.K. (2016) A multivariate analysis integrating ecological, socioeconomic and physical characteristics to investigate urban forest cover and plant diversity in Beijing, China. *Ecological Indicators* 60: 921–929.

Watts C.W. and Dexter A.R. (1997) The influence of organic matter in reducing the destabilization of soil by simulated tillage. *Soil and Tillage Research* 42(4): 253–275.

WCCD (2016) *World Council on City Data: ISO 37120*. Available at http://open.dataforcities. org/ (accessed 4 August 2017).

Wiedmann T.O., Schandl H., Lenzen M., Moran D., Suh S., West J. and Kanemoto K. (2015) The material footprint of nations. *Proceedings of the National Academy of Sciences* 112(20): 6271–6276.

Wilson E.O. (1984) *Biophilia*. Harvard University Press, Cambridge, MA.

Yang X., Li Z. and Webster C. (2016) Estimating the mediate effect of privately green space on the relationship between urban public green space and property value: Evidence from Shanghai, China. *Land Use Policy* 54: 439–447.

Yuan F. and Bauer M.E. (2007) Comparison of impervious surface area and normalized difference vegetation index as indicators of surface urban heat island effects in Landsat imagery. *Remote Sensing of Environment* 106(3): 375–386.

Zhao J., Ouyang Z., Xu W., Zheng H. and Meng X. (2010) Sampling adequacy estimation for plant species composition by accumulation curves: A case study of urban vegetation in Beijing, China. *Landscape and Urban Planning* 95(3): 113–121.

4

PLANNING FOR PROTECTION

Promoting pest suppressing urban landscapes through habitat management

Dieter F. Hochuli and Caragh G. Threlfall

Introduction

The maintenance of healthy vegetation in urban landscapes is essential for creating sustainable cities. In addition to providing habitat for numerous biodiversity groups, city trees have been shown to play a vital role in supporting human health through the effects on well-being as well as moderating the effects of pollution and urban heat (Willis and Petrokofsky 2017). The emerging paradigm of promoting green infrastructure in cities has created enormous opportunities for ecologists to identify how best to manage greenspace in cities to promote multiple benefits for humans and nature.

Green infrastructure and greenspace in cities take many forms (Taylor and Hochuli 2017). Street trees are seen as vital for promoting everyday human interactions with nature (Willis and Petrokofsky 2017), while urban forests often represent relatively extensive examples of ecosystems providing what some see as a more authentic nature experience. In both cases, these socio-ecological systems are inevitably subjected to a range of anthropogenic pressures that affect abiotic and biotic drivers of vegetation health. These pressures can manifest in declines of plant quality and health, particularly under elevated levels of insect herbivory.

Insect herbivore outbreaks and subsequent pest damage on vegetation raises significant challenges for land managers in urban ecosystems (Raupp *et al.* 2010). Chemical interventions, typically used to control insect pests, are often seen as a last resort, owing to the potential impacts on humans and the broader environment. The potential to exploit *beneficial* biodiversity, animals known to prey upon pest species, is an ideal way to promote biodiversity-friendly nature-based solutions in urban systems.

In this chapter, we outline the potential to manage urban landscapes to promote *beneficial* animals who play a vital ecological role regulating insect herbivores, with a view to creating pest-supressing landscapes. In doing this, we draw from the rich history of using insects in agricultural systems for this purpose, identifying potential strategies for promoting missing components of biodiversity in urban ecosystems, and link these in a framework where landscape management at multiple scales can foster the restoration and enhancement of lost ecological interactions.

Ecosystem services and biodiversity

In all landscapes, biodiversity underpins ecosystem function and supports all human activities (Millennium Ecosystem Assessment 2005). However, our understanding of how to better manage the urban environment to maintain ecosystem function is limited. Many critical ecosystem functions are provided specifically by organisms that move across the landscape (Lundberg and Moberg 2003), including plant pollination, seed dispersal and pest insect control. These mobile-agent-based ecosystem services (MABES) (Kremen *et al.* 2007) are estimated to be worth US \$10–100 billion to agriculture and forestry (Costanza *et al.* 1997), and are provided by organisms such as insects and birds (Kremen *et al.* 2007). Animal-mediated services benefit people via their influence on food production, plant regeneration and vegetation health, on which other services depend (Kremen *et al.* 2007). These services are produced locally, and the organisms responsible for them come from, and are influenced by, the surrounding landscape (Lundberg and Moberg 2003). Hence, the design of cities almost certainly influences both the organism and resulting services. This is of concern as the loss of MABES leads to declines in the reproductive success of plants (Biesmeijer *et al.* 2006), reduced crop yield (Kremen *et al.* 2002; Losey and Vaughan 2006), and increased use of chemicals and pesticides (Boyles *et al.* 2011). These services are extremely costly to restore if lost (Losey and Vaughan 2006), and hence warrant substantial research attention in urban landscapes.

Regulation of herbivores by predators and parasites in urban systems

The role of predators and parasites in regulating prey populations is central to many fundamental principles in ecology. The *green world hypothesis*, proposed in 1960, identifies the suite of factors that limit herbivores, highlighting the potential for predators to reduce herbivore abundances and create ideal environments for plants to thrive (Hairston *et al.* 1960).

A rich body of experimental work in ecology shows how *top-down* control by predators of prey can lead to enormous shifts in the extent of insect herbivory caused by insectivorous predators as diverse as spiders (Schmitz 2008), lizards (Dial and Roughgarden 1995), birds (Marquis and Whelan 1994) and bats (Boyles *et al.* 2011).

Wasp and fly parasitoids also play a major role in regulating insect herbivores (Bianchi *et al.* 2008). As such, the success of beneficial biodiversity in controlling insect pests on plants is unsurprising.

In urban systems, a key issue influencing the health of vegetation is the disruption of ecological interactions through the loss of species, particularly those from higher trophic levels (Faeth *et al.* 2005). Predators and parasitoids are generally more susceptible to disturbance (Ryall and Fahrig 2006), and their loss is seen to be a major contributor to elevated herbivory in degraded and disturbed systems (Dwyer *et al.* 2004; Raupp *et al.* 2010).

In Figure 4.1 we outline a framework by which these interactions in urban systems can be disrupted by the loss of higher trophic levels, and identify ways in which effects can be assessed. We argue that pervasive anthropogenic stressors expose urban ecosystems to extensive biodiversity loss and subsequent disruptions to ecological interactions. We also outline the potential benefits of restoring these ecological interactions through targeting the ecological needs of higher trophic levels and providing them across multiple spatial scales.

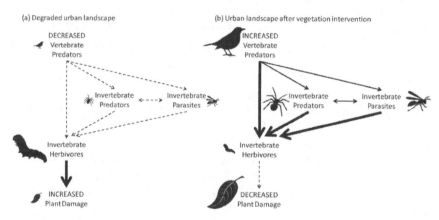

FIGURE 4.1 Simplified food web showing (a) disruptions to trophic relationships in urban systems and (b) potential to restore and manage trophic relationships using vegetation interventions. Dotted lines indicate weakening of interactions while thickness of bold lines indicates increasingly stronger trophic cascades. This framework identifies the recognised effects of landscape degradation that lead to declines in the abundances of both beneficial vertebrates and invertebrates, and how these effects of urbanisation combine to reduce top-down regulation, leading to increased herbivore abundance and plant damage. Hence, although the loss of vertebrate predators in urban system shown in (a) may have a minor effect on invertebrate predators and parasitoids, these effects are weakened by the declines in the abundances of invertebrate predators and parasitoids caused by landscape degradation.

Elevated herbivory in urban environments

Overviews examining the ecology of herbivorous arthropods in cities reveal evidence of high levels of herbivory in multiple contexts (Raupp et al. 2010). Concerns over defoliation in urban ecosystems have been raised at multiple spatial scales; from street trees, to dominant trees in parks and across entire urban forests (Hochuli et al. 2004; Christie and Hochuli 2005; Dale and Frank 2014). The aesthetic and ecological impact of damage to foliage in urban ecosystems often raises considerable challenges for land managers, who have limited options for controlling the extent of damage.

There are likely to be multiple causal drivers for these losses, relating to the gamut of urban stressors impacting upon these systems. For example, there is strong evidence from street trees in Raleigh, North Carolina that urban heat island effects affect insect–plant interactions, through differential effects on herbivore and parasitoids (Meineke et al. 2013; Dale and Frank 2017). Elevated temperatures favour some insect pests (Meineke et al. 2013), through promoting rapid development and acclimation of certain pest species while their parasitoids lag behind. Species loss, driven by habitat modification, is also likely to be a significant driver of trophic disruptions. For example, there is considerable evidence that elevated herbivory on dominant trees in urban forests in Sydney, Australia (Christie and Hochuli 2005) is a function of the loss of insectivorous predators, causing the loss of top-down control. Importantly, the reduction of top-down control appears to be a factor of a general loss across multiple insectivorous guilds (Christie et al. 2010).

However, the pattern of elevated herbivory in urban ecosystems is not universal. A recent multi-city study showed reduced levels of herbivory on birch trees in 16 cities throughout Europe, linked to elevated predation by ants and birds in cities (Kozlov et al. 2017) and reveals the importance of system-specific approaches. While most evidence points to significant losses of species in cities, particularly from higher trophic levels, there are clearly circumstances where more resilient urban species can persist and thrive, playing important ecological roles (McDonnell and Hahs 2015).

The potential to modify habitats to support components of biodiversity playing important ecological roles, such as pest suppression, creates not only an opportunity to address species loss in these systems, but also to combat disruptions to ecological interactions and trophic regulation.

Landscape interventions and beneficial biodiversity

The long history of successful landscape management in agriculture and horticulture to enhance populations of beneficial biodiversity provides a roadmap for urban ecologists and land managers (Landis et al. 2000; Gurr et al. 2016). Strategies to promote populations of parasitoids, insect predators and vertebrate predators remain central to enhancing yields in a multitude of systems. The key to the success of these strategies is an understanding of the ecological requirements of the desired species, understanding their trophic and regulatory roles through experimental manipulations, and integrating these into landscape scale experiments testing the effectiveness of management approaches on both patterns of distribution and ecological process. Further, in non–urban landscapes, taking a landscape-scale approach has generated an understanding of the landscape attributes that support ecosystem function, and enabled the development of spatially explicit management plans to retain biodiversity and ecosystem services (Kremen and Cowling 2005). Yet this approach has rarely been taken in urban landscapes.

Many of the effective management approaches encouraging beneficial biodiversity are deceptively simple. Providing non–crop floral resources has been shown to enhance the conservation of beneficial arthropods (Balzan *et al.* 2016), often leading to reduced pest densities and levels of damage (Tschumi *et al.* 2015). Plant species loss has also been shown to shift trophic structures of ecosystems by decreasing the diversity of arthropods in them (Haddad *et al.* 2009), confirming the importance of maintaining diverse plant communities as well. The identity and composition of nearby habitat is also a major driver for populations of beneficial biodiversity, with areas that offer more structural complexity and diversity, typically providing a source for beneficial fauna (Den Belder *et al.* 2002). These patterns are not limited to beneficial invertebrates, with invertebrate-feeding birds in agricultural systems regularly responding to landscape attributes such as habitat structure (Atkinson *et al.* 2004) and plant identity (Barrett *et al.* 2008). The benefits, historical and current, of biological control in agro–ecosystems are manifold (Crowder and Jabbour 2014), and reflect the potential for developing pest-suppressing landscapes in other systems where plant health needs to be managed.

Concerns over the environmental effects of pesticides and impacts on city dwellers make this promotion of beneficial biodiversity strategies ideal options for land managers in urban systems. Importantly, urban ecologists do not need to reinvent the wheel to exploit the potential to use predators and parasitoids to limit damage. The evidence required to demonstrate strategies that work in agricultural systems is generally twofold; the initial support that target fauna respond to the management actions and critically, that this response has an ecological impact (e.g. reduced damage on target flora). The recent recognition that many of the effects of landscape composition are scale dependent in agricultural systems (Martin *et al.* 2016) is particularly salient for urban ecologists. Understanding how beneficial biodiversity in urban landscapes responds to landscape modification requires similar evidence, always considering a third dimension; scale-specific frameworks.

Scale and ecological interactions in urban ecosystems

The response of biodiversity to increased urban density and habitat alteration has been studied at a variety of spatial scales. Biodiversity response at the scale of an entire city is governed by different social, cultural, economic and biological drivers than biodiversity response at smaller spatial scales, such as at the neighbourhood or parcel scale (Aronson *et al.* 2017). Indeed, at the lot or parcel scale, property owners may make very different decisions about vegetation arrangement, structure and composition

(Aronson *et al.* 2017) which may in turn lead to different biodiversity responses at larger spatial scales. Household scale management choices may in fact conflict with city-wide goals for conservation or biodiversity management, and empirical studies are only just beginning to understand the importance of decisions made at different scales on ecological processes (Aronson *et al.* 2017). It makes sense then, to discuss the impact of habitat management on animal communities in relation to the scales relevant to urban landscapes and the different actors or decision makers present at each scale, including parcel, street-scales and city-wide scales.

Promoting beneficial invertebrates in urban environments

While vertebrate predators typically dominate the literature on trophic cascades (Shurin *et al.* 2002), the role of invertebrate predators and parasitoids in regulating pests is undisputed (Gurr *et al.* 2016). These groups of animals often persist in cities, with many being surprisingly resilient to the pressures of urbanisation (Christie and Hochuli 2009; Christie *et al.* 2010). However, they have also been shown to respond to a range of landscape attributes and habitat characteristics.

Importantly, while the desirability to use natural habitat to enhance biological control of pests makes it an attractive option for land managers, it is also important to recognise that these methods are not a panacea for pest suppression (Tscharntke *et al.* 2016). The key to successful habitat interventions remains identifying the mechanisms driving change for different components of the beneficial fauna.

Local-scale habitat management

The mechanisms underlying the accumulation of natural enemies in complex habitats are poorly known, although meta-analyses suggest that a multitude of factors such as reduced intraguild predation, more effective prey capture, higher prey abundance and access to alternative resources (alternative prey, pollen or nectar) may have an impact (Langellotto and Denno 2004). The results of this meta-analysis supported the view that bottom–up resources mediated top–down impacts on herbivores. This provides encouragement that manipulations of habitat complexity in agroecosystems can support pest suppression by enhancing habitat for assemblages of natural enemies (Langellotto and Denno 2004).

Strong associations in urban systems between habitat structure and abundance and diversity have been reported for spiders (Otoshi *et al.* 2015), parasitoids (Christie and Hochuli 2009; Bennett and Gratton 2012) and general insect predators (Pereira-Peixoto *et al.* 2016). The built environment can also have significant effects on biodiversity. Simple physical barriers, such as walls, have been found to reduce connectivity in urban habitats, such that colonisation by herbivores is reduced in the presence of walls (Peralta *et al.* 2011). Green roofs may also supplement populations of beneficial insects, although the ability of herbivores and their parasitoids to colonise these habitats may be compromised, with some components of the parasitoids unable to exploit these new environments (Quispe and Fenoglio 2015).

The ecology of different species may also be affected by scale. A meta-analysis of landscape scale studies conducted in agroecosystems (Chaplin-Kramer *et al.* 2011) demonstrated that natural enemy communities positively respond to complexity at landscape scales, but that generalist and specialist enemies had differing responses. Landscape complexity refers to the amount and diversity of natural or non-crop habitats in the areas surrounding farms and fields, and was found to be an important driver of abundance and diversity of enemy communities, including invertebrate predators and parasitoids. Interestingly, specialist enemies were shown to respond to habitat complexity at finer spatial scales than generalists, suggesting that management targeting the control of a specific pest is likely to be more effective if the identity and ecology of its natural enemy is taken into account.

Augmenting flower strips in crop land and vineyards has been shown to be an effective and simple method for promoting beneficial insects at fine scales (Wratten *et al.* 2007). Recent investigations of urban green spaces suggest that invertebrate predators including beetles and bugs respond to changes in habitat structure at fine spatial scales, where the presence of native vegetation and complex understorey habitats increases the probability of occurrence for many species within these taxa (Mata *et al.* 2017; Threlfall *et al.* 2017). Similarly, predatory flies and spiders also respond positively to the presence of uncompacted soils and high flower diversity within small patches of habitat in urban green spaces (Gardiner *et al.* 2014).

Street-scale – patch scale

While street trees and associated green infrastructure in cities make significant contributions to human interests (Hartig and Kahn 2016; Willis and Petrokofsky 2017), their effects on invertebrate biodiversity may be limited owing to the highly disturbed context they occur in. The anthropogenic stressors often encountered by street trees (e.g. urban heat island effects, nutrient supplementation, pollution, modified drainage) limit their growth and development (Quigley 2004) and do not have immediate parallels in agroecosystems. The effects of these stressors on responses by biodiversity remain largely unexplored and a key area for future research in urban systems.

While street trees are often used to mitigate the effects of urban heat islands, they may also be significantly affected by urban heat. For example, street trees may suffer extensive damage associated with changes in thermal conditions (Meineke *et al.* 2013). These increases in damage observed may be linked to differential responses to urban heat among herbivores and natural enemies (Dale and Frank 2017). These effects may occur at community scales, with spiders responding to thermal characteristics in these systems rather than prey availability (Meineke *et al.* 2017).

Furthermore, the ecology of street trees in the urban matrix may influence patch dynamics and biodiversity in urban systems beyond those predicted from simple assessments using traits of larger scale patches and fragmentation frameworks from landscape ecology (Rossetti *et al.* 2017). Street trees can influence patch dynamics, and play a role as stepping stones or even sources for beneficial biodiversity, providing significant opportunity for small scale local habitat interventions to have larger scale impacts.

Landscape-scale management (city-wide)

The extensive body of literature showing invertebrate responses to urbanisation at landscape scales reveals distinctive general patterns that are tied to taxonomic group, functional group and species identity (Hochuli *et al.* 2009). The reliance on descriptive gradient-based approaches generally reveals declines in biodiversity in cities, despite some species surviving and thriving in urban landscapes (McDonnell and Hahs 2015). Urbanisation can affect the trophic structure of invertebrate assemblages at landscape scales (Christie *et al.* 2010), although ecological interactions in cities can be decoupled at multiple scales (Nelson and Forbes 2014).

While the diverse literature examining invertebrate responses to urbanisation at landscape scales reveals a wide array of responses, three key factors are regularly identified as important drivers of diversity: (1) extent of habitat, (2) quality of habitat and (3) spatial relation to other habitats (primary connectedness and isolation) all contribute to the capacity of an area to support beneficial biodiversity.

The extent of habitat available is seen as a major driver of biodiversity composition, with larger, relatively undisturbed areas providing important habitat for a range of taxa, such as wasps, ants and beetles in cities (Gibb and Hochuli 2002, Hochuli *et al.* 2009). The traits of this available habitat (such as vegetation composition and structural complexity) are also key drivers of diversity (Christie and Hochuli 2009; Bennett and Gratton 2012; Philpott *et al.* 2014; Ossola *et al.* 2015). These attributes match those described as important for natural enemies in agroecosystems (Gurr *et al.* 2016), further reinforcing the benefits of looking to these agroecosystems for management options.

While successful whole city approaches are likely to focus on preserving high quality remnants or restoring degraded landscapes to promote beneficial invertebrate biodiversity (Hochuli *et al.* 2009), gains may also be made outside these conventional paths. Enhancing the potential for the urban matrix to support beneficial biodiversity is also a key option for managing beneficial biodiversity locally to generate city-wide outcomes. While highly modified areas in urban ecosystems are typically seen as inhospitable to biodiversity, the potential to exploit green infrastructure, such as green walls, roofs (Hartig and Kahn 2016) and gardens (Philpott and Bichier 2017), offers enormous potential to enhance beneficial biodiversity at landscape scales.

Promoting beneficial vertebrates in urban environments

Due to their density in many urban areas, birds and bats are likely to contribute significantly to pest reduction. However, this service is poorly recognised and may be reduced in the future if biodiversity continues to decline with increasing urban intensification. Commonly reported garden pests include white cabbage moth, and many species of sap-sucking bugs that are recognised prey in the diet of a variety of bird and bat species. This suggests that the retention of these vertebrate insectivores could greatly benefit urban vegetation.

Local-scale habitat management

Habitat management that alters vegetation structure directly and indirectly influences foraging resource abundance and availability for birds. Further, the response of birds to the management of vegetation is largely guild dependant. For example, Atkinson *et al.* (2004) suggest that in rural grassland systems, birds that feed on soil invertebrates are generally tolerant to or positively respond to management that maintains shorter swards of grasses and patches of bare soil, as this enables foraging access. Conversely, this type of grassland management negatively influences birds that feed on resources provided by short grassy vegetation (seeds, sap-sucking invertebrates). Agricultural fields that contain swards of vegetation at multiple heights, with moderate levels of cutting and organic fertiliser input may result in increased seed production and invertebrate abundance, however taller patches of vegetation may restrict accessibility of these food resources to certain bird species. In the case of insectivorous birds, studies of revegetated patches in agricultural landscapes suggest that planting native forbs may positively influence this group, but that increasing vegetation structure alone through tree planting is unlikely to be sufficient (Barrett *et al.* 2008). Hence, management approaches aimed at maintaining or restoring habitat for birds would benefit from taking a species or guild specific approach.

The retention or provision of nesting resources is also critical to enhancing habitat for predatory vertebrates. Increases in structural complexity in urban green spaces may improve nesting habitat provision for many guilds of birds, including insectivores (Threlfall *et al.* 2016b), in addition to supporting a greater number of insectivorous bat species (Threlfall *et al.* 2017). Many bird and bat species nest in tree hollows and cavities provided by large, old trees, which are a scarce resource in human impacted landscapes (Le Roux *et al.* 2014; Threlfall *et al.* 2016a). The provision of nest boxes is one strategy employed to conserve cavity dependent insectivorous species in many landscapes. The addition of nest boxes can lead to significant increases in the densities of insectivorous birds, and has been demonstrated to increase rates of invertebrate control in vineyards via increased predation by birds (Jedlicka *et al.* 2011). Similarly, the provision of nest boxes designed for bats in Mediterranean rice paddies led to a decline of the rice borer moth (*Chilo supressalis*) shortly after box installation and occupancy by bats (Puig-Montserrat *et al.* 2015). However, the use of nest boxes in an urban context has not been widely evaluated.

Street-scale – patch scale

The presence of street-tree networks may go some way to mitigating the negative impact of increased urban density on mobile vertebrates. In a study of bats in the West Midlands metropolitan county, a highly urbanised region of the UK, Hale *et al.* (2012) found that connected street tree networks provided habitat connectivity for certain species of bats, even in densely built up areas. They concluded that some species of bats may respond positively to connected tree networks at scales between 100 and 500 m, especially if critical features such as water, roosting habitat and foraging habitat are located in close proximity to high density tree networks. Similarly, street-tree networks have been demonstrated to be important for the retention of diverse bird communities, where suburbs with greater than 30 per cent composition of native street trees retain greater bird species richness (Ikin *et al.* 2013).

Landscape-scale management (city-wide)

Biodiversity returns on landscape-scale efforts to revegetate or restore habitat have most often been assessed in agricultural settings. Munro *et al.* (2007) reviewed the literature on faunal response to revegetation efforts in Australia and concluded that fauna response was positively related to revegetation that occurs in large and complex patches. Fauna, in particular many species of birds, also respond to the age of plantings, and their level of connectivity (Munro *et al.* 2007; Barrett *et al.* 2008). At a city-wide scale; revegetation or restoration efforts are often not co-ordinated, as urban land parcels are most often managed in an individual way (Aronson *et al.* 2017) or public land is managed by different local authorities. Further, efforts at city-wide scales are made more complex by the considerable economic costs associated with dedicating land to conservation, and questions around the efficacy of such actions. Instead, there may be opportunity to restore habitat via management of the urban matrix, by making the matrix in between vegetated patches more permeable to the movements of native species (e.g. by increasing the density of street-tree canopy or encouraging wildlife gardening in backyards). However, few urban studies on the topic published to date provide conclusive findings.

Mobile animals such as birds and bats not only rely on the presence of critical habitat at fine scales; they also rely on the retention of suitable habitat at city-wide scales. For example, in Sydney, Australia, in parts of the city where the cover of vegetation fell below 30 per cent the number of species that occurred significantly declined (Threlfall *et al.* 2012). Similarly, in Melbourne, Australia, a decrease in bat species occurrence was observed when tree cover fell below 20 per cent in moderately urbanised areas (Caryl *et al.* 2016). Further, in areas of high housing density, additional tree cover did not increase the probability of occurrence of patch and edge dependant bat species (Caryl *et al.* 2016), suggesting that highly urbanised areas cannot support these species, regardless of the level of tree cover. Other factors limiting the occurrence of diverse bat communities in city centres include the lack of suitable roosting sites, especially for species that do not use man-made structures, negative impacts of artificial night light and potentially increased predation pressure. Despite these threats, other more common and resilient species still occur throughout highly urbanised areas, although these species do not include the gleaning and urban sensitive bats (Jung and Threlfall 2016) that play a major role in insect regulation in non-urban landscapes (Williams-Guillen *et al.* 2008).

Promoting ecological interactions in cities

While there is a substantial and growing body of work examining how different components of biodiversity respond to habitat manipulations, there is an urgent need to couple these with assessments of the factors affecting trophic relationships and ecological interactions in cities. There is surprisingly little work coupling the shifts in the composition of biodiversity with measures of ecological function, with an overwhelming majority of work focusing on assessing the former. Put simply, we often have a strong understanding of which traits may influence the occurrence of biodiversity in urban ecosystems (McDonnell and Hahs 2015) but have little evidence to evaluate the consequences (Faeth et al. 2005). The reliance on static measures of success in urban ecology (occurrence and loss of certain species) does not address the critical question of what the ecological significance of these is.

Particularly lacking is experimental work testing how predation rates on insect pests ultimately have an effect on levels of herbivory. Introductions of sentinel pests to urban gardens in California revealed strong effects of both landscape and local drivers of predation services in these systems (Philpott and Bichier 2017). However, previous work fails to survey fauna across multiple trophic levels (Faeth et al. 2005; Christie et al. 2010). Integrating the breadth of faunal responses to urbanisation and coupling this finer scale experimental work assessing trophic cascades will address the significant gap in our understanding of the putative interactions in urban food webs and the implications for urban pest management.

Conclusions

Owing to the complex nature of urban ecosystems it is impossible to advocate a simple *one size fits all* approach to promoting pest-supressing landscapes. The nature of different cities means that the advice must be tailored to the ecology of that city, acknowledging it in the context of historical and current disturbances. The contrasting patterns in herbivory driven by the responses of insectivorous birds to urbanisation in Europe (Kozlov et al. 2017) and Australia (Christie and Hochuli 2005) show the importance of developing city-specific solutions when creating pest-suppressing landscapes.

Even at a finer scale within cities, vegetation is likely to be subjected to a range of different pressures and occur in vastly different contexts (Aronson et al. 2017). A species of plant in an urban forest will encounter vastly different challenges compared to when it exists as a street tree. When identifying local and landscape drivers of ecological interactions in cities it will be central to identify processes that are scale- and context-dependent (Aronson et al. 2017, Martin et al. 2016).

Nevertheless, there are strong general principles that may guide the search for best practices for developing pest-supressing landscapes in cities. Many of these are derived from integrating the success of these approaches in agroecosystems, with the growing literature understanding how biodiversity responds to urbanisation.

In Figure 4.2 we present a schematic that outlines simple measures that could guide principles for promoting elements of biodiversity that play a role in supressing pests. These are:

- promoting native ground cover and limiting mowing;
- adding habitat structure in understorey (providing shelter and habitat);
- adding floral diversity for food and shelter (i.e. availability of flowering plants);
- adding plant diversity;
- planting with a long term perspective, accommodating current needs and future climates;
- retaining habitat traits that provide natural nesting sites such as hollow-bearing trees, and logs and rocks; and
- adding artificial roosting and nesting habitat when necessary.

These actions are by definition local and can be applied at fine scales. As such, they may fail to address the challenges of creating pest-suppressing landscapes across cities. A meta-analysis of the local and landscape scale factors that influence urban biodiversity, including birds and invertebrates, suggested that retention of vegetation at city-wide scales is one of the most important factors driving the response of biodiversity to urbanisation (Beninde *et al.* 2015). Importantly though, despite this recognition that the quantity of vegetation in the landscape is critical, clearly the quality of this vegetation at finer spatial scales must be considered due to its impact on the abundance and diversity of taxa across multiple trophic levels.

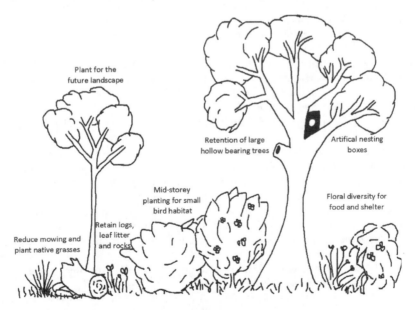

FIGURE 4.2 Habitat interventions to encourage beneficial biodiversity in urban ecosystems.

Maintaining healthy vegetation in cities in the face of the suite of threats faced by urban greenspace requires both local and city-wide approaches. The challenge for urban ecologists is to integrate assessments of biodiversity responses to interventions with experiments testing their ecological impacts. In that sense, we see each vegetation intervention as an experiment waiting to be done. In a rapidly urbanising world where the maintenance of greenspace is seen as critical for both biodiversity and human well-being (Taylor and Hochuli 2015), there are enormous opportunities for urban ecologists to use experimental approaches to develop a knowledge base that allows fine scale vegetation interventions to target critical species, creating pest-suppressing landscapes.

Acknowledgements

We thank Margot Law for drawing Figures 4.1 and 4.2 and initial literature searches. CT is supported by the Clean Air and Urban Landscapes Hub, which is funded by the Australian Government's National Environmental Science Programme.

References

Aronson, M.F.J., Lepczyk C.A., Evans K.L., Goddard M.A., Lerman S.B., MacIvor J.S., Nilon C.H. and Vargo T. (2017) Biodiversity in the city: Key challenges for urban green space management. *Frontiers in Ecology and the Environment* 15: 189–196.

Atkinson P.W., Buckingham D. and Morris A.J. (2004) What factors determine where invertebrate-feeding birds forage in dry agricultural grasslands? *Ibis* 146: 99–107.

Balzan M.V., Bocci G. and Moonen A.-C. (2016) Utilisation of plant functional diversity in wildflower strips for the delivery of multiple agroecosystem services. *Entomologia Experimentalis et Applicata* 158: 304–319.

Barrett G.W., Freudenberger D., Drew A., Stol J., Nicholls A.O. and Cawsey E.M. (2008) Colonisation of native tree and shrub plantings by woodland birds in an agricultural landscape. *Wildlife Research* 35: 19–32.

Beninde J., Veith M. and Hochkirch A. (2015) Biodiversity in cities needs space: A meta-analysis of factors determining intra-urban biodiversity variation. *Ecology Letters* 18: 581–592.

Bennett A.B. and Gratton C. (2012) Local and landscape scale variables impact parasitoid assemblages across an urbanization gradient. *Landscape and Urban Planning* 104: 26–33.

Bianchi F., Goedhart P. and Baveco J. (2008) Enhanced pest control in cabbage crops near forest in The Netherlands. *Landscape Ecology* 23: 595–602.

Biesmeijer J.C., Roberts S.P.M., Reemer M., Ohlemüller R., Edwards M., Peeters T., Schaffers A.P., Potts S.G., Kleukers R., Thomas C.D., Settele J. and Kunin W.E. (2006) Parallel declines in pollinators and insect-pollinated plants in Britain and the Netherlands. *Science* 313: 351–354.

Boyles J.G., Cryan P.M., McCracken G.F. and Kunz T.H. (2011) Economic importance of bats in agriculture. *Science* 332: 41–42.

Caryl F.M., Lumsden L.F., van der Ree R. and Wintle B.A. (2016) Functional responses of insectivorous bats to increasing housing density support 'land-sparing' rather than 'land-sharing' urban growth strategies. *Journal of Applied Ecology* 53: 191–201.

Chaplin-Kramer, R., O'Rourke M.E., Blitzer E.J. and Kremen C. (2011) A meta-analysis of crop pest and natural enemy response to landscape complexity. Ecology Letters 14: 922–932.

Christie F.J. and Hochuli D.F. (2005) Elevated levels of herbivory in urban landscapes: Are declines in tree health more than an edge effect? *Ecology and Society* 10(1): 10.

Christie F.J. and Hochuli D.F. (2009) Responses of wasp communities to urbanization: Effects on community resilience and species diversity. *Journal of Insect Conservation* 13: 213–221.

Christie F.J., Cassis G. and Hochuli D.F. (2010) Urbanization affects the trophic structure of arboreal arthropod communities. *Urban Ecosystems* 13: 169–180.

Costanza R., d'Arge R., de Groot R., Farber S., Hannon B. and Grasso M. (1997) The value of the world's ecosystem services and natural capital. *Nature* 387: 253–260.

Crowder D.W. and Jabbour R. (2014) Relationships between biodiversity and biological control in agroecosystems: Current status and future challenges. *Biological Control* 75: 8–17.

Dale A.G. and Frank S.D. (2014) The effects of urban warming on herbivore abundance and street tree condition. *PLoS ONE* 9: e102996.

Dale A.G. and Frank S.D. (2017) Warming and drought combine to increase pest insect fitness on urban trees. *PLoS ONE* 12: e0173844.

Den Belder E., Elderson J., Van Den Brink W. and Schelling G. (2002) Effect of woodlots on thrips density in leek fields: A landscape analysis. *Agriculture, Ecosystems & Environment* 91: 139–145.

Dial R. and Roughgarden J. (1995) Experimental removal of insectivores from rainforest canopy: Direct and indirect effects. *Ecology* 76: 1821–1834.

Dwyer G., Dushoff J. and Yee S.H. (2004) The combined effects of pathogens and predators on insect outbreaks. *Nature* 430: 341–345.

Faeth S.H., Warren P.S., Shochat E. and Marussich W.A. (2005) Trophic dynamics in urban communities. *BioScience* 55: 399–407.

Gardiner M.M., Prajzner S.P., Burkman C.E., Albro S. and Grewal P.S. (2014) Vacant land conversion to community gardens: Influences on generalist arthropod predators and biocontrol services in urban greenspaces. *Urban Ecosystems* 17: 101–122.

Gibb H. and Hochuli D.F. (2002) Habitat fragmentation in an urban environment: Large and small fragments support different arthropod assemblages. *Biological Conservation* 106: 91–100.

Gurr G.M., Wratten S.D., Landis D.A. and You M. (2016) Habitat management to suppress pest populations: Progress and prospects. *Annual Review of Entomology* 62: 91–109.

Haddad N.M., Crutsinger G.M., Gross K., Haarstad J., Knops J.M.H. and Tilman D. (2009) Plant species loss decreases arthropod diversity and shifts trophic structure. *Ecology Letters* 121029–1039.

Hairston N.G., Smith F.E. and Slobodkin L.B. (1960) Community structure, population control, and competition. *American Naturalist* 94: 421–425.

Hale J.D., Fairbrass A.J., Matthews T.J. and Sadler J.P. (2012) Habitat composition and connectivity predicts bat presence and activity at foraging sites in a large UK conurbation. *PLoS ONE* 7: e33300.

Hartig T. and Kahn P.H. (2016) Living in cities, naturally. *Science* 352: 938–940.

Hochuli D.F., Gibb H., Burrows S.E., Christie F.J., Lunney D. and Burgin S. (2004) Ecology of Sydney's urban fragments: Has fragmentation taken the sting out of insect herbivory? *Urban Wildlife*, ch. 9: 63–69. Available at http://publications.rzsnsw.org.au/doi/abs/10.7882/FS.2004.082 (accessed 4 August 2017).

Hochuli D.F., Lomov B. and Christie F.J. (2009) Invertebrate biodiversity in urban land-scapes: Assessing remnant habitat and its restoration. In McDonnell, M.J., Hahs, A.K. and Breuste J.H. (eds), *Ecology of Cities and Towns: A Comparative Approach*, pp. 215–232. Cambridge University Press, Cambridge.

Ikin K., Knight E., Lindenmayer D.B., Fischer J. and Manning A.D. (2013) The influence of native versus exotic streetscape vegetation on the spatial distribution of birds in sub-urbs and reserves. *Diversity and Distributions* 19: 294–306.

Jedlicka J.A., Greenberg R. and Letourneau D.K. (2011) Avian conservation practices strengthen ecosystem services in California vineyards. *PLoS ONE* 6: 1–8.

Jung K. and Threlfall C.G. (2016) Urbanisation and its effects on bats: A global meta-analysis. In Voigt C.C. and Kingston T. (eds), *Bats in the Anthropocene: Conservation of Bats in a Changing World*, pp. 13–33. Springer International Publishing, Cham.

Kozlov M.V., Lanta V., Zverev V., Rainio K., Kunavin M.A. and Zvereva E.L. (2017) Decreased losses of woody plant foliage to insects in large urban areas are explained by bird predation. *Global Change Biology*, 23: 4354–4364.

Kremen C. and Cowling R. (2005) Managing ecosystem services: What do we need to know about their ecology? *Ecology Letters* 8: 468–479.

Kremen C., Williams N.M., Aizen M.A., Gemmill-Herren B., LeBuhn G., Minckley R., Packer L., Potts S.G., Roulston T.A., Steffan-Dewenter I., Vázquez D.P., Winfree R., Adams L., Crone E.E., Greenleaf S.S., Keitt T.H., Klein A.-M., Regetz J. and Ricketts T.H. (2007) Pollination and other ecosystem services produced by mobile organisms: A conceptual framework for the effects of land-use change. *Ecology Letters* 10: 299–314.

Kremen C., Williams N.M. and Thorp R.W. (2002) Crop pollination from native bees at risk from agricultural intensification. *Proceedings of the National Academy of Sciences* 99: 16812–16816.

Landis D.A., Wratten S.D. and Gurr G.M. (2000) Habitat management to conserve natural enemies of arthropod pests in agriculture. *Annual Review of Entomology* 45: 175–201.

Langellotto G.A. and Denno R.F. (2004) Responses of invertebrate natural enemies to complex-structured habitats: A meta-analytical synthesis. *Oecologia* 139: 1–10.

Le Roux D.S., Ikin K., Lindenmayer D.B., Manning A.D. and Gibbons P. (2014) The future of large old trees in urban landscapes. *PLoS ONE* 9: 1–11.

Losey J.E. and Vaughan M. (2006) The economic value of ecological services provided by insects. *Bioscience* 56: 311–323.

Lundberg J. and Moberg F. (2003) Mobile link organisms and ecosystem functioning: Implications for ecosystem resilience and management. *Ecosystems* 6: 0087–0098.

McDonnell M.J. and Hahs A.K. (2015) Adaptation and adaptedness of organisms to urban environments. *Annual Review of Ecology, Evolution, and Systematics* 46: 261–280.

Marquis R.J. and Whelan C.J. (1994) Insectivorous birds increase growth of white oak through consumption of leaf-chewing insects. *Ecology* 75: 2007–2014.

Martin E.A., Seo B., Park C.-R., Reineking B. and Steffan-Dewenter I. (2016) Scale-dependent effects of landscape composition and configuration on natural enemy diversity, crop herbivory, and yields. *Ecological Applications* 26: 448–462.

Mata L., Threlfall C., Williams N., Hahs A., Malipatil M., Stork N. and Livesley S. (2017) Conserving herbivorous and predatory insects in urban green spaces. *Scientific Reports* 7: 40970.

Meineke E.K.,Dunn R.R., Sexton J.O. and Frank S.D. (2013) Urban warming drives insect pest abundance on street trees. *PLoS ONE* 8: e59687.

Meineke E.K., Holmquist A.J., Wimp G.M. and Frank S.D. (2017) Changes in spider com-munity composition are associated with urban temperature, not herbivore abundance. *Journal of Urban Ecology* 3: 1–8.

Millennium Ecosystem Assessment (2005) *Ecosystems and Human Well-Being: Synthesis.* Island Press, Washington DC.

Munro N.T., Lindenmayer D.B. and Fischer J. (2007) Faunal response to revegetation in agricultural areas of Australia: A review. *Ecological Management & Restoration* 8: 199–207.

Nelson A.E. and Forbes A.A. (2014) Urban land use decouples plant-herbivore-parasitoid interactions at multiple spatial scales. *PLoS ONE* 9: e102127.

Ossola A., Nash M., Christie F., Hahs A. and Livesley S.J. (2015) Urban habitat complexity affects species richness but not environmental filtering of morphologically-diverse ants. *PeerJ* 3: e1356.

Otoshi M.D., Bichier P. and Philpott S.M. (2015) Local and landscape correlates of spider activity density and species richness in Urban Gardens. *Environmental Entomology* 44: 1043–1051.

Peralta G., Fenoglio M.S. and Salvo A. (2011) Physical barriers and corridors in urban habitats affect colonisation and parasitism rates of a specialist leaf miner. *Ecological Entomology* 36: 673–679.

Pereira-Peixoto M.H., Pufal G., Staab M., Feitosa Martins C. and Klein A.-M. (2016) Diversity and specificity of host-natural enemy interactions in an urban-rural interface. *Ecological Entomology* 41: 241–252.

Philpott S.M. and Bichier P. (2017) Local and landscape drivers of predation services in urban gardens. *Ecological Applications* 37: 966–976.

Philpott S.M., Cotton J., Bichier P., Friedrich R.L., Moorhead L.C., Uno S. and Valdez (2014) Local and landscape drivers of arthropod abundance, richness, and trophic composition in urban habitats. *Urban Ecosystems* 17: 513–532.

Puig-Montserrat X., Torre I., López-Baucells A., Guerrieri E., Monti M.M., Ràfols-García R., Ferrer X., Gisbert D. and Flaquer C. (2015) Pest control service provided by bats in Mediterranean rice paddies: Linking agroecosystems structure to ecological functions. *Mammalian Biology – Zeitschrift für Säugetierkunde* 80: 237–245.

Quigley M.F. (2004) Street trees and rural conspecifics: Will long-lived trees reach full size in urban conditions? *Urban Ecosystems* 7: 29–39.

Quispe I. and Fenoglio M.S. (2015) Host–parasitoid interactions on urban roofs: An experimental evaluation to determine plant patch colonisation and resource exploitation. *Insect Conservation and Diversity* 8: 474–483.

Raupp M.J., Shrewsbury P.M. and Herms D.A. (2010) Ecology of herbivorous arthropods in urban landscapes. *Annual Review of Entomology* 55: 19–38.

Rossetti M.R., Tscharntke T., Aguilar R. and Batáry P. (2017) Responses of insect herbivores and herbivory to habitat fragmentation: A hierarchical meta-analysis. *Ecology Letters* 20: 264–272.

Ryall K.L. and Fahrig L. (2006) Response of predators to loss and fragmentation of prey habitat: A review of theory. *Ecology* 87: 1086–1093.

Schmitz O.J. (2008) Effects of predator hunting mode on grassland ecosystem function. *Science* 319: 952–954.

Shurin J.B., Borer E.T., Seabloom E.W., Anderson K., Blanchette C.A., Broitman B., Cooper S.D. and Halpern B.S. (2002) A cross-ecosystem comparison of the strength of trophic cascades. *Ecology Letters* 5: 785–791.

Taylor L. and Hochuli D. (2015) Creating better cities: How biodiversity and ecosystem functioning enhance urban residents' wellbeing. *Urban Ecosystems* 18: 747–762.

Taylor L. and Hochuli D. (2017) Defining greenspace: Multiple uses across multiple disciplines. *Landscape and Urban Planning* 158: 25–38.

Threlfall C.G., Law B. and Banks P.B. (2012) Sensitivity of insectivorous bats to urbanization: Implications for suburban conservation planning. *Biological Conservation* 146: 41–52.

Threlfall C.G., Ossola A., Hahs A.K., Williams N.S.G., Wilson L. and Livesley S.J. (2016a) Variation in vegetation structure and composition across urban green space types. *Frontiers in Ecology and Evolution* 4: 66.

Threlfall C.G., Williams N.S.G., Hahs A.H. and Livesley S.J. (2016b). Approaches to urban vegetation management and the impacts on urban bird and bat assemblages. *Landscape and Urban Planning* 153: 28–39.

Threlfall C.G., Mata L., Mackie J.A., Hahs A.K., Stork N.E., Williams N.S. and Livesley S.J. (2017) Increasing biodiversity in urban green spaces through simple vegetation interventions. *Journal of Applied Ecology* (in press), doi:10.1111/1365-2664.12876.

Tscharntke T., Karp D.S., Chaplin-Kramer R., Batáry P., DeClerck F., Gratton C., Hunt, Ives A., Jonsson M. and Larsen A. (2016) When natural habitat fails to enhance biological pest control: Five hypotheses. *Biological Conservation* 204: 449–458.

Tschumi M., Albrecht M., Entling M.H. and Jacot K. (2015) High effectiveness of tailored flower strips in reducing pests and crop plant damage. *Proceedings of the Royal Society B: Biological Sciences* 282: 20151369.

Williams-Guillen K., Perfecto I. and Vandermeer J. (2008) Bats limit insects in a neotropical agroforestry system. *Science* 320(5872): 70.

Willis K.J. and Petrokofsky G. (2017) The natural capital of city trees. *Science* 356: 374–376.

Wratten S.D., Hochuli D.F., Gurr G.M., Tylianakis J., Scarratt S.L. (2007) Conservation, biodiversity, and integrated pest management. In Kogan M. and Jepson P. (eds), *Perspectives in Ecological Theory and Integrated Pest Management*, pp. 223–245. Cambridge University Press, Cambridge, UK.

5

URBAN AGRICULTURE

An opportunity for biodiversity and food provision in urban landscapes

Brenda B. Lin and Monika H. Egerer

Introduction

In the face of urbanization and growing human populations, urban green spaces play an important role in harbouring biodiversity within city landscapes, maintaining trophic interactions and food web stability, and providing ecosystem services to urban residents (Goddard *et al.* 2010). Urban agriculture – the production of food and livestock in urban areas – is an important feature of urban green infrastructure. These agro-ecosystems can be considered islands of high biodiversity in the urban landscape as they generally contain an abundance of species of plants, birds, and arthropods in comparison to the surrounding urban landscape matrix (Goddard *et al.* 2010). This is especially true for beneficial insects such as pollinators (e.g., Ahrné *et al.* 2009) and natural enemies (e.g., Bennett and Gratton 2012). Organisms such as bees, flies, butterflies, spiders, and beetles provide key ecological functions through pollination and pest predation, which lead to ecosystem services that increase plant and crop production in both rural (Losey and Vaughan 2006) and urban agricultural systems (Lin *et al.* 2015). Urban agriculture is thus a key space for biodiversity and ecosystem service provisioning that can increase local food production and urban food security and access (Smit *et al.* 1996). As urban populations grow across the world, urban agriculture is becoming ever-more important for its socio-economic implications such as increased food security and nutrition (Alaimo *et al.* 2008), its significant role in biodiversity conservation and in urban ecology (Lin *et al.* 2015), and overall integrating multifunctionality into densely populated urban landscapes (Lovell 2010).

What is urban agriculture?

Urban agriculture (UA) is defined as the production of crop and livestock goods within cities and towns (Zezza and Tasciotti 2010), and it is generally integrated into the local urban economic and ecological systems (Mougeot 2010). UA often includes peri-urban agricultural areas around cities and towns, which may provide products and services to the local urban population (Mougeot 2010). Urban agriculture activities are diverse and can include the cultivation of vegetables, medicinal plants, spices, mushrooms, fruit trees and other productive plants, as well as keeping livestock for eggs, milk, meat, wool, and other products (Lovell 2010). The different types of UA contribute to the edible landscape in a range of community types and provide a broad array of services based on community needs and desires (McLain *et al.* 2012). This can include spaces

TABLE 5.1 Different typologies and descriptions of UA systems to highlight the diversity of urban farming.

Type	*Description*	*References*
Community or allotment gardens	Represent small-scale, highly patchy, and qualitatively rich (vegetatively complex and species rich) agro-ecosystems that are usually located in urban or semi-urban areas for food production.	Colding *et al.* 2006
Private gardens	Primarily located in suburban areas and may be the most prevalent form of urban agriculture in cities. For example, privately owned gardens cover an estimated 22–27% of the total urban area in the UK, 36% in New Zealand, and 19.5% in Dayton, Ohio, USA.	Loram *et al.* 2007; Mathieu *et al.* 2007
Easement gardens	Gardens often regulated by the local government but located within private or community properties. Urban easements are established with the purpose of improving water quality and erosion control, but they can include a wide array of biodiversity, including food plants, depending on management type. Gardening on verges may also be done as a form of *guerrilla gardening* where local communities garden on small patches of soil when few unpaved spaces are available.	Hunter and Hunter 2008; Hunter and Brown 2012
Rooftop gardens or green roofs	Any vegetation established on the roof of a building and can be used to improve insulation, create local habitat, provide decorative amenity, and cultivate food plants.	Whittinghill and Rowe 2012
Urban orchards	Tree-based food production systems that can be owned and run privately or by the community. Increasingly, schools and hospitals are establishing fruit trees that provide crops, erosion control, shade, and wildlife habitat, and producing food for the local community.	Drescher *et al.* 2006
Peri-urban agriculture	Usually exists at the outskirts of cities that largely serve the needs of the nearby urban population. Typically, these are functional agricultural systems that include a large variety of activities and diversification approaches, and contribute to environmental, social and economic functions.	Zasada 2011

such as private gardens (household area privately cultivated), community gardens (areas collectively cultivated), allotment gardens (parcelled areas individually cultivated), and peri-urban farms (production-focused systems) (Table 5.1). UA systems are highly heterogeneous in size, form, and function and can be found in different types of urban green spaces. This diversity is based on some important factors including land tenure, management, production type and scale of production.

Many UA systems fit into more than one category. For example, both private gardens and community gardens may exist as rooftop gardens, and orchards may exist within community gardens. They may be cultivated by an individual owner or by a community. The various types of UA that exist are important toward providing the planned vegetative diversity necessary to support other associated biodiversity within cities (Figure 5.1).

Urban agriculture as important areas of food provision in cities

Urban agriculture is increasingly supported within and around cities due to food security concerns. Several US cities contain *food deserts*, where access to fresh produce is limited due to reduced proximity to markets, financial constraints, or inadequate transportation (ver Ploeg *et al.* 2009; Thomas 2010). For example, assessments of the Oakland, CA food system have underscored that affordability is the most important factor that influences where low-income residents shop for food (Wooten 2008), and residents' limited access to transportation to grocery stores is another fundamental constraint to accessing healthy food (Treuhaft *et al.* 2009).

UA has rapidly increased in developing countries all over the world, especially since the 2008 increase of global food prices (FAO 2014). In many African nations, for example, the percentage of low-income urban population participating in UA has grown from 20 per cent in the 1980s to about 70 per cent in the 2000s (Bryld 2003). This is because UA can be very productive, providing an estimated 15–20 per cent of the global food supply (Smit *et al.* 1996; Hodgson *et al.* 2011). For example, UA provides 60 per cent of the vegetables and 90 per cent of the eggs consumed by residents in Shanghai, 47 per cent of the produce in urban Bulgaria, 60 per cent of vegetables in Cuba, and 90–100 per cent of the leafy vegetables in poor households of Harare, Zimbabwe (Lovell 2010).

Additionally, as urban crop cultivation can also provide significant dietary contributions, communities around the world are using it to improve the health of urban residents. Many successful UA programmes have increased the food security of local residents. For example, New York City's (NYC) *Green Thumb* has become the largest community gardening programme in the US, with more than 600 gardens that support 20,000 urban residents located in ethnically and culturally diverse neighbourhoods where a wide range of community members cultivate and manage the gardens (Lovell 2010). Ongoing expansion in Detroit's urban gardening scene is expected to produce 31 per cent of the vegetables and 17 per cent of the fruits currently consumed by city residents on just 100–350 ha of land (Colasanti and Hamm 2010). Private gardens also contribute significantly to local food production and

FIGURE 5.1 The diverse forms of urban agriculture: a) rooftop restaurant garden in San Cristóbal, Chiapas, MX; b) campus farm at the University of California, Santa Cruz, CA; c) City Parks & Recreation garden in San Jose, CA; d) Non-profit garden with *adopted* beds in Salinas, CA.

food security. A study in Chicago showed that the food production area of home gardens was almost threefold that of community gardens. This suggests that home food gardens can contribute significantly to enhancing community food sovereignty (Taylor and Lovell 2012).

Urban agriculture can support high levels of biodiversity

Urban agriculture is an increasingly important urban green space in which to support and enhance urban biodiversity. If designed carefully and deliberately, urban gardens can support high levels of biodiversity and ecosystem services, which in turn allow for more resilient food production systems. Thus, it is important to evaluate the design and management factors that maximize biodiversity and ecosystem services coming from gardens so that communities can be best served by these spaces.

Urban agriculture in the context of urban biodiversity research

Urbanization has been shown to be a force of biotic homogenization where species assemblages across cities become more similar because the similar challenges of the built environment across cities (habitat fragmentation, pollutants, etc.) select for species that can survive and thrive in these systems (e.g. pigeons) (McKinney 2006). This type of selection fundamentally changes patterns of regional biodiversity (species distribution, competition, etc.) (Schwartz *et al.* 2006) as the urban matrix increasingly dominates the landscape. However, research in urban systems over the past decades has studied how local and landscape structure can support ecologically significant biodiversity, such as insects, plants and birds, among others (Beninde *et al.* 2015). For example, studies on urban insect pollinator community richness, pollination and pollinator meta-populations have unveiled how organisms interact within local habitats and respond to the urban matrix (Jha and Kremen 2013; Lowenstein *et al.* 2014).

While research has studied ecological interactions in urban gardens, many studies often focus broadly on all green spaces (e.g., parks, hedgerows, cemeteries) (Andersson *et al.* 2007), thereby obscuring the specific ecological dynamics and importance of urban agriculture in its contributions to urban biodiversity. As part of the green infrastructure of urban landscapes, UA can exhibit a wide breadth of biotic diversity and provide critical resources to species sensitive to detrimental side effects of urbanization, thus combating the homogenization effect. In the following sections, we describe how specific local and landscape habitat characteristics, as well as human characteristics of urban neighbourhoods, influence biodiversity and associated environmental and cultural functions such as food provisioning.

Local and landscape factors affect biodiversity in urban agriculture

Both local factors (i.e., habitat characteristics) and landscape factors (i.e., surrounding landscape features) affect the degree to which agroecosystems contribute to biodiversity and ecosystem services such as food provisioning (Altieri 1999; Tscharntke *et al.* 2005), and can be applied to urban environments (Angold *et al.* 2006). Local factors include vegetative diversity, abundance of crops and flowering plants, and soil management practices. Landscape factors can include landscape connectivity, landscape diversity within a reference area, and the position along a rural to urban gradient. Yet, our understanding of how these factors interact at both levels with one another to affect biodiversity and ecosystem function is still relatively limited in the urban context (Angold *et al.* 2006; Matteson and Langellotto 2010).

Local factors: management and environmental heterogeneity

Habitat management in urban agriculture by vegetative and soil management supports local biodiversity, ecological interactions among organisms (i.e., food webs), and food production for people. Habitat size and quality is a key driver of urban biodiversity (Beninde et al. 2015) such as beneficial insects (Angold et al. 2006; Pardee and Philpott 2014). Urban vegetation management can strategically aim to increase habitat quality for ecological communities in urban agroecosystems because plants provide a *template* for ecological community formation and species interactions (Faeth et al. 2011). Urban gardens can harbour rich floral and ornamental plant communities, providing nectar and trophic resources to support beneficial insect populations and the overall species diversity (Colding et al. 2006), as well as high diversity of flowering vegetable and fruit crops, reflective of the cultural diversity of community gardeners (Baker 2004).

It is thus little surprise that larger urban gardens with greater flower and plant abundance and diversity have been related with increased beneficial insect abundance and species richness. Bee community richness increases with urban garden size (Frankie et al. 2005), and floral and plant abundance and richness (Frankie et al. 2005; Matteson and Langellotto 2010; Pardee and Philpott 2014). Smith et al. (2006) also found that solitary bee diversity in urban agroecosystems was positively correlated with the presence of certain vegetative components, such as trees, and overall structural complexity. Additionally, Bennett and Gratton (2012) found that high Hymenopteran parasitoid abundance was best explained by increased flower diversity within urban green spaces, which is a finding that has implications for pest control services in urban gardens.

In response to growing concerns over bee populations and pollination services and public popularity, studies have focused on urban bee pollinator responses to local vegetation composition in gardens with implications for ecosystem function and services. In comparison to other urban green spaces, urban community gardens with high ornamental and flower diversity are critical habitats for urban bees, and most studies have found strong correlations between local vegetation characteristics and pollinators. Colding et al. (2006) assessed community and domestic gardens in Stockholm, Sweden, and found that community gardens had a high abundance and diversity of flowering plants (over 400 species) that in turn supported a high abundance and diversity of urban pollinators. Urban gardens have been found to harbour both increased abundance and diversity of bumble bees in response to greater flower presence and richness (Andersson et al. 2007, Ahrné et al. 2009), and overall bee species diversity in response to crop plant and ornamental diversity (Matteson et al. 2008). Additionally, Matteson and Langellotto (2010) found strong relationships between butterfly and bee diversity and local floral resources as well as wild areas within urban gardens in New York City. Similarly, Chicago's neighbourhood bee abundance and richness increased as a response to floral diversity and also to human presence, suggesting pollination services are mediated by residents planting a diversity of flowering plants (Lowenstein et al. 2014).

The effect of native plants within urban gardens on beneficial insect populations and ecosystem function is still debatable. Matteson and Langellotto (2011) studied the impact of native plant additions on bee, butterfly, and predatory wasp species richness in urban community gardens previously dominated by exotic flowers in New York City. They found that increasing native plants in urban gardens did not attract more insect visitors or contribute to visitor diversity or abundance. However, bee presence and abundance has been linked to native plants that are present in urban and suburban areas (Frankie 2005), and increasing native plantings of flowers has indeed been shown to strongly enhance native bee populations in urban community gardens from 5 to 31 species in three years (Pawelek *et al.* 2009). Along with those findings, Pardee and Philpott (2014) found that native plants provide floral resources for native bee populations in a resource-poor urban landscape in Toledo, Ohio, and stress the importance of native plants in urban gardens to support urban arthropod diversity and abundance, and ecosystem functions such as pollination. Interestingly, they found that bee community composition was significantly different between native and non-native gardens, providing evidence that plant community composition influences pollinator community composition.

These differences in response to native plantings may be due to several factors. First, exotic species can provide greater nectar resources for butterflies and bees, thus attracting and fostering biodiversity. Second, a significant increase in pollinator abundance and diversity may require larger and more diverse native plant additions, as well as longer sampling periods. Matteson and Langellotto (2011) were limited to a 16-month study, while Pawalek *et al.* (2009) observed results of planting manipulations over a three year duration. Third, urban context and the degree of land-use disturbance intensity and frequency may be a driver of observed differences (Pardee and Philpott 2014). In this comparison, New York City's insect biodiversity may be comprised of more generalist feeders (Matteson and Langellotto 2011) or other life history characteristics adapted to more intensely managed and disturbed land. Future assessments will require a standardized research methodology in both scale and sampling duration to determine the role of native planting manipulations in influencing biodiversity and abundance of beneficial insect populations, and should compare and contrast floral nectar resource availability. These research efforts could greatly benefit from long term partnerships between researchers and the community of practice, including garden organizational leadership and participants as well as city parks and recreation services. In sum, garden practitioners and local to city-wide management practices can support urban insect populations important for ecosystem function and services in urban agriculture for increased food provision.

Landscape factors: structure and connectivity in the urban landscape

Urban agriculture is distributed across a complex urban landscape. Land-use configuration can enhance or block ecological functions within local ecosystems as built environments generally result in increased impervious cover and fragmentation of urban green spaces, leading to habitat area decline and a reduction in species diversity (McKinney 2006). Fragmentation has been shown to negatively impact urban insect and arthropod populations such as pollinators (Cane *et al.* 2006).

The declines in urban bees is likely a function of increased impervious cover, as increasing impervious cover decreases habitat area, and bee foraging and dispersal movement (Jha and Kremen 2013). Bumble bee diversity follows this trend and was found to decrease in response to increasing landscape impervious cover (Ahrné *et al.* 2009). Increased mobility and dispersal of functional insects is a result of landscape matrix permeability, which is influenced by degree of urban development (or amount of impervious cover), overall complexity of the landscape, and habitat connectivity. Lin and Fuller (2013) equate urban landscape mosaics with agricultural landscapes, as they can be similar in both their homogenization and intensity of land use. In rural agricultural landscapes, landscape-level intensification can have negative impacts on beneficial insects, and pest control and pollination (Tscharntke *et al.* 2005). This has been shown for urban natural enemy abundances; parasitoid wasp abundance declines as the percent of urban green decreases and impervious cover increases (Bennett and Gratton 2012). Pollinator population abundance also declines in response to urban development intensity (Jha and Kremen 2013). Thus, natural areas can provide source populations and resources to urban agricultural systems to increase the abundance and diversity of functional organisms such as bees (Hernandez *et al.* 2009).

To confront negative impacts of fragmentation and urbanization, connectivity and the creation of *green corridors* have been proposed to enhance abundance, diversity and ecosystem function within and among urban green spaces (Rudd *et al.* 2002). Consistent with this approach, various urban conservation programmes have focused on creating green pathways via planting flowers and native grasses in utility easements, hedgerows, riparian corridors and backyard gardens, to support urban wildlife and add ecological value to cities (Rudd *et al.* 2002). This supports UA biodiversity while also allowing UA to be a critical node of these green corridors.

Few have rigorously assessed the efficacy of enhanced connectivity for urban agriculture insect diversity and agroecosystem function. Rudd *et al.* (2002) show that urban gardens can facilitate functional connectivity of urban green spaces, and suggest gardening as a tool to enhance regional habitat quality. Colding *et al.* (2006) demonstrate the importance of garden connectivity within a fragmented and heavily developed urban landscape, with evidence that allotment garden networks support urban metapopulations of native bees by facilitating movement and enhancing pollination function. Thus, urban agriculture has the potential to significantly contribute to the overall green space connectivity and should be integrated into conservation and planning models to increase urban biodiversity and maintain ecosystem services. Further, local management practices can increase habitat quality within gardens and result in high abundance and species richness of insect populations. High quality habitats with high local biodiversity and ecosystem functions can potentially have a *spillover* effect across a landscape when high degrees of landscape connectivity and permeability exist. Thus, connecting urban gardens to existing forms of green corridors at the landscape level can potentially enhance habitat configuration and permeability for mobile biodiversity, and in turn increase insect species abundance, richness, and dispersal to support urban populations and ecosystem functions such as pollination.

Social systems can affect urban agriculture management and food provision

Urban systems are heavily influenced by the environmental conditions (e.g. built environment, changes in climate, changes in water flows), as well as by social conditions (e.g., planning, finance, community attitudes and desires). These conditions also exist within urban gardens, where plant selection, management, and soil preparation are highly affected by the social complexity of networks, organizations, knowledge flows, and the power dynamics of gardens and their communities. The nuances of these aspects within the community can drive the motivation, values, and interactions of individuals to influence the management of these spaces and the associated biodiversity and ecosystem service generation (Andersson *et al.* 2007).

The biodiversity of urban agriculture is often *infused* with the human diversity and pluralism of metropoles (Baker 2004). In an examination of the community gardening movement in Toronto, Canada, Baker (2004) found that elderly gardeners use specific agricultural techniques developed as farmers in rural China to grow culturally appropriate foods for themselves. In central California, Corlett and colleagues (2003) found that urban garden biodiversity and ethnobotany reflects the origins of urban farmers: nearly all of the 59 species of plants reported by Hmong farmers had a cited use in Southeast Asian literature for food, seasoning or medicine. These examples demonstrate that ethno-cultural diversity is reflected in agricultural practices and agrobiodiversity in urban agriculture.

Uneven patterns in urban development leave behind very heterogeneous landscapes and heterogeneous patterns in socio-economic gradients (Swan *et al.* 2011). Social and economic variation as a result of income inequality can drive urban plant species diversity to influence associated biotic communities (Hope *et al.* 2008). Termed the *luxury effect*, as urban neighbourhood wealth (median family income) increases, plant species diversity can also increase in urban areas (Hope *et al.* 2008). The luxury effect can have bottom-up influences on higher trophic levels within the ecological community, such as urban park and neighbourhood avian diversity (Kinzig *et al.* 2005). This suggests that residents within neighbourhoods of lower socio-economic status are experiencing inequitable access to biodiversity-rich urban environments and suggests an inequitable distribution of ecosystem services. Further, this trend is also documented in urban agriculture: in a recent study Clarke and Jenerette (2015) examined the relationships among indicators of economic wealth, human ethnic diversity and plant crop diversity in Los Angeles community gardens. They found that ornamental flower species diversity and abundance in urban community gardens per garden plot significantly increased with neighbourhood wealth. The authors also found significant trends among dominant gardeners' ethnicity and species composition: gardens categorized into predominant ethnicities (Non-immigrant, African-American, Asian and

Hispanic) were self-similar in their species composition of food crops and ornamental species compared to gardens of different ethnic groups. This suggests that pollination services may be different considering the importance of ornamental flower diversity and composition for pollinator guilds.

The correlative results call for more information on individual motivations, knowledge and values, and social networks that may influence garden biodiversity management. Urban agriculture leads to not just increased urban food production, but increased social interactions; in allotment gardens, gardeners often exchange ecological and cultural knowledge and experience with other gardeners (Saldivar-Tanaka and Krasny 2004) to influence personal management practices (Andersson *et al.* 2007). Thus, there can be a tangible biodiversity spillover not only from plot to plot, but a spillover in the ecological knowledge and learning from gardener to gardener.

Beyond food provision: the socio-ecological benefits and challenges of urban agriculture

Urban agricultural is common across continents with urban gardens covering hundreds of hectares in Amsterdam, Montreal, Beijing, and Barcelona, among many other cities (reviewed in Lovell 2010), and such green spaces serve many environmental and social uses for urban citizens. UA is regarded as an important feature for the long-term support of urban systems at global scale (Barthel and Isendahl 2013), and thus critical to the sustainability and resilience of cities. Additionally, with many benefits to cities, urban policy, and development have been increasingly adopted to introduce and maintain such systems (McClintock and Cooper 2010). However, some challenges are associated with agricultural systems in cities with many interests competing for land use.

We have tangentially described many of the benefits related to urban agriculture. Researchers and popular media have highlighted the social–ecological benefits and multifunctionality of urban agriculture (Lovell 2010). Urban agriculture is associated with forms of civic agriculture and food justice, community development and social networks, and urban greening and recreation. These benefits are a reason why urban agriculture has been encouraged within alternative agri-food movements. First, civic agriculture emphasizes the localization of food production, and embeds the agri-food system within the social, economic, and ecological systems of a place (Lyson 2004). As a form of civic agriculture, urban agriculture increases urban food security, fresh food access, public health, and food sovereignty. Household participation in community gardens increases fresh fruit and vegetable intake among participants as observed in Flint, MI where gardeners were 3.5 times more likely to get five servings of fruits and vegetables daily (Alaimo *et al.* 2008). Further, urban agriculture may offer an opportunity for farmers to grow culturally appropriate, high quality and diverse foods unavailable at the store (Baker 2004), and to utilize their agricultural knowledge to define their own diets (Minkoff-Zern 2012).

Urban agriculture offers a suite of other social benefits that may be more important than the actual food growing. A space for daily socializing, community bonding, education, and special events are all well documented benefits of community gardens

(Saldivar-Tanaka and Krasny 2004). In New York City, community gardeners value gardens as spaces for reading, writing, and studying in addition to skill-based workshops to learn about farming/cultivation practices, cooking and nutrition (Saldivar-Tanaka and Krasny 2004). Further, urban community gardens are also sites of community and citizenship where women form community based on ethnicity and knowledge sharing, and the shared experience of adapting to a new country (Corlett *et al.* 2003).

However, urban agriculture projects also face many challenges as a community and social movement. Projects confront logistical barriers such as land access, soil contamination from previous industrialization, and lack of water (Guitart *et al.* 2012). Projects also confront structural barriers such as accessing and maintaining property rights (Irazabal and Punja 2009). Thus, urban agriculture projects struggle to be sustainable long-term efforts due to compounding challenges related to land security, gentrification, capital, and human resources. Further, many projects may not address social justice or issues of race and inequality present in the alternative agri-food movement (Reynolds and Cohen 2016). In sum, urban agriculture can be considered spaces of ecosystem services and ecological wealth (e.g., food, pollination, biodiversity), yet simultaneously spaces of ecosystem disservices (e.g., invasive species or nutrient run-off) and social injustices (e.g., inequitable distribution of resources and environmental pollutants). The interplay between services and disservices has challenged researchers to understand the ecological and social complexities of these systems, and how social–ecological interactions spill over across the urban landscape.

Supporting urban agriculture and its contributions to city life

To understand the myriad of benefits that UA provides, it is imperative that we protect and maintain these green spaces in rapidly densifying cities. Recent studies have revealed relationships among biodiversity, ecosystem functions and services and several local, landscape, and social factors, showing that:

- biodiversity in UA systems is highly human managed. As vegetation structural complexity and composition in urban agroecosystems is a result of local management, insect diversity for ecosystem functioning can be human-mediated;
- landscape heterogeneity, and increased permeability via increased green space can affect insect species movement, and local biodiversity and community composition, affecting the biodiversity and ecosystem function of UA systems.

Though many of the studies highlighted here have examined local and landscape factors in urban systems, few have looked at both specifically in the urban agroecosystem context to evaluate their relative importance on both abundance and species richness of functional species. Supporting research in urban agriculture that assesses what local, landscape and social factors affect specific UA ecosystem functions is necessary to develop policies that promote UA systems (Figure 5.2). Integrative and multiple approaches can determine how landscape connectivity of urban gardens can promote species mobility and population numbers, ecosystem

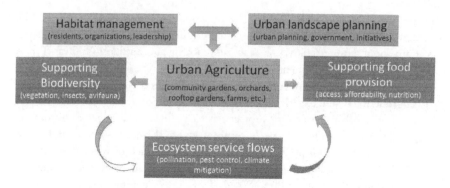

FIGURE 5.2 Urban agriculture involves a diversity of stakeholders that in turn influence how urban agriculture supports the conservation of urban biodiversity, the flow of ecosystem services, and the cultivation of food benefits.

service multiplicity across space, and how to facilitate urban garden networks to increase ecosystem function.

While urban pollinators and their services are well studied, there is a gap in our understanding of other ecosystem functions and services in urban agroecosystems. Urban research that mirrors rural agricultural studies can provide information on how an urban matrix and an agricultural matrix compare in functional insect responses to habitat. Additionally, urban gardens may provide other ecosystem services that are not insect-related, including carbon sequestration and storage and water conservation by vegetation and soil (Davies *et al.* 2011). Thus, urban agroecosystems not only harbour biodiversity essential for ecosystem functions, but can be beneficial for climate regulation and resource conservation.

The influence of management and incorporation of stakeholders in urban gardens is often not studied in tandem with ecological research, despite the fact that local management by residents and organizations has wide implications for increasing ecosystem resilience and service provisioning (Colding *et al.* 2006). A large portion of urban studies end by addressing the role of citizen management in mediating resource-providing habitat to set the stage for ecosystem functions (Andersson *et al.* 2007; Lowenstein *et al.* 2014). For example, creating an urban bee-friendly habitat should involve the participation of community gardens to develop conservation initiatives that focus on creating a garden vegetation structure shown to be correlated to diverse bee communities (Hernandez *et al.* 2009). Thus, a high priority for urban ecologists should be to communicate with and involve stakeholders early in the research process both to develop context relevant questions, and to provide applicable information to garden practitioners and urban planners concerned with creating supportive agroecosystems. Urban gardens, farms, and forms of agroecosystems have the opportunity to be areas of biodiversity conservation and to increase local food security. Connecting ecological research on urban agriculture with practitioners can facilitate realistic conservation efforts of urban biodiversity and ecosystem processes that benefit urban social and ecological communities.

References

Ahrné K., Bengtsson J. and Elmqvist T. (2009) Bumble bees (*Bombus spp*) along a gradient of increasing urbanization. *PLoS ONE* 4(5): e5574.

Alaimo K., Packnett E., Miles R.A. and Kruger D.J. (2008) Fruit and vegetable intake among urban community gardeners. *Journal of Nutrition Education and Behavior* 40: 94–101.

Altieri M. (1999) The ecological role of biodiversity in agroecosystems. *Agriculture, Ecosystems, and Environment* 74: 19–31.

Andersson E., Barthel S. and Ahrné K. (2007) Measuring social–ecological dynamics behind the generation of ecosystem services. *Ecological Applications* 17(5): 1267–1278.

Angold P.G., Sadler J.P., Hill M.O., Pullin A., Rushton S., Austin K., Small E., Wood B., Wadsworth R., Sanderson R. and Thompson K. (2006) Biodiversity in urban habitat patches. *Science of the Total Environment* 360: 196–204.

Baker L. (2004) Tending cultural landscapes and food citizenship in Toronto's community gardens. *Geographic Review* 94: 305–325.

Barthel S. and Isendahl C. (2013) Urban gardens, agriculture, and water management: Sources of resilience for long-term food security in cities. *Ecological Economics* 86: 224–234.

Beninde J., Veith M. and Hochkirch A. (2015) Biodiversity in cities needs space: A meta-analysis of factors determining intra-urban biodiversity variation. *Ecology Letters* 18: 581–592.

Bennett A.B. and Gratton C. (2012) Local and landscape scale variables impact parasitoid assemblages across an urbanization gradient. *Landscape and Urban Planning* 104(1): 26–33.

Bryld E. (2003) Potentials, problems, and policy implications for urban agriculture in developing countries. *Agriculture and Human Values* 20(1): 79–86.

Cane J.H., Minckley R.L., Kervin L.J., Roulston T.H. and Williams N.M. (2006) Complex responses within a desert bee guild (Hymenoptera: Apiformes) to urban habitat fragmentation. *Ecological Applications* 16(2): 632–644.

Clarke L.W. and Jenerette G.D. (2015) Biodiversity and direct ecosystem service regulation in the community gardens of Los Angeles, CA. *Landscape Ecology* 30: 637.

Colasanti K.A. and Hamm M.W. (2010) Assessing the local food supply capacity of Detroit, Michigan. *Journal of Agriculture, Food Systems, and Community Development* 1(2): 41.

Colding J., Lundberg J. and Folke C. (2006) Incorporating green-area user groups in urban ecosystem management. *AMBIO: A Journal of the Human Environment* 35(5): 237–244.

Corlett J.L., Dean E.A. and Grivetti L.E. (2003) Hmong gardens: Botanical diversity in an urban setting. *Economic Botany* 57(3): 365–379.

Davies Z.G., Edmondson J.L., Heinemeyer A., Leake J.R. and Gaston K.J. (2011) Mapping an urban ecosystem service: Quantifying above-ground carbon storage at a city-wide scale. *Journal of Applied Ecology* 48: 1125–1135.

Drescher A., Holmer R. and Iaquinta D. (2006) *Urban Homegardens and Allotment Gardens for Sustainable Livelihoods: Management Strategies and Institutional Environments, Tropical Homegardens*, pp. 317–338. Springer, The Netherlands.

Faeth S.H., Bang C. and Saari S. (2011) Urban biodiversity: Patterns and mechanisms. *Annals of the New York Academy of Sciences* 1223(1): 69–81.

FAO (2014) *Growing Green Cities in Latin America and the Caribbean*. Food and Agriculture Organization of the United Nations, Rome, Italy.

Frankie G.W., Thorp R.W., Schindler M., Ertter B. and Rizzardi M. (2005) Ecological patterns of bees and their host ornamental flowers in two northern California cities. *Journal of the Kansas Entomological Society* 78(3): 227–246.

Goddard M.A., Dougill A.J. and Benton T.G. (2010) Scaling up from gardens: Biodiversity conservation in urban environments. *Trends in Ecology and Evolution* 25(2): 90–98.

Guitart, D., Pickering, C. & Byrne, J., 2012. Past results and future directions in urban community gardens research. *Urban Forestry and Urban Greening* 11(4): 364–373.

Hernandez J.L., Frankie G.W. and Thorp R.W. (2009) Ecology of urban bees: A review of current knowledge and directions for future study. *Cities and the Environment* 2(1): 3.

Hodgson K., Campbell M.C. and Bailkey M. (2011) *Urban Agriculture: Growing Healthy, Sustainable Places*. American Planning Association, Washington, DC.

Hope D., Gries C., Zhu W., Fagan W.F., Redman C.L., Grimm N.B., Nelson A.L., Martin C. and Kinzig A. (2008) Socioeconomics drive urban plant diversity. *Proceedings of the National Academy of Sciences* 100(15): 339–347.

Hunter M.C.R.and Hunter M.D. (2008) Designing for conservation of insects in the built environment. *Insect Conservation and Diversity* 1(4): 189–196.

Hunter M.C.R. and Brown D.G. (2012) Spatial contagion: Gardening along the street in residential neighborhoods. *Landscape and Urban Planning* 105(4): 407–416.

Irazabal C. and Punja A. (2009) Cultivating just planning and legal institutions: A critical assessment of the South Central Farm struggle in Los Angeles. *Journal of Urban Affairs* 31(1): 1–23.

Jha S. and Kremen C. (2013) Bumble bee foraging in response to landscape heterogeneity. *Proceedings of the National Academy of Sciences* 110(2): 555–558.

Kinzig A.P., Warren P., Martin C., Hope D. and Katti M. (2005) The effects of human socioeconomic status and cultural characteristics on urban patterns of biodiversity. *Ecology and Society* 10(1): 23.

Lin B.B. and Fuller R.A. (2013) Sharing or sparing? How should we grow the world's cities? *Journal of Applied Ecology* 50(5): 1161–1168.

Lin B.B., Philpott S.M. and Jha S. (2015) The future of urban agriculture and biodiversity ecosystem-services: challenges and next steps. *Basic and Applied Ecology* 16(3): 189–201.

Loram A., Tratalos J., Warren P. and Gaston K.J. (2007) Urban domestic gardens (X): The extent & structure of the resource in five major cities. *Landscape Ecology* 22(4): 601–615.

Losey J.E. and Vaughan M. (2006) The economic value of ecological services provided by insects. *BioScience* 56(4): 311–323.

Lovell S.T. (2010) Multifunctional urban agriculture for sustainable land use planning in the United States. *Sustainability* 2(8): 2499–2522.

Lowenstein D.M. Matteson K.C. Xiao I., Silva A.M. and Minor E.S. (2014) Humans, bees, and pollination services in the city: The case of Chicago, IL (USA). *Biodiversity and Conservation* 23(11): 2857–2874.

Lyson T.A. (2004) *Civic Agriculture: Reconnecting Farm, Food, and Community*. Tufts University Press, Medford, MA.

Mathieu R., Freeman C. and Aryal J. (2007) Mapping private gardens in urban areas using object-oriented techniques and very high-resolution satellite imagery. *Landscape and Urban Planning* 81(3): 179–192.

Matteson K.C. and Langellotto G.A. (2010) Determinants of inner city butterfly and bee species richness. *Urban Ecosystems*, 13: 333–347.

Matteson K.C. and Langellotto G.A. (2011) Small scale additions of native plants fail to increase beneficial insect richness in urban gardens. *Insect Conservation and Diversity* 4: 89–98.

Matteson K.C., Ascher J.S. and Langellotto G.A. (2008) Bee richness and abundance in New York City urban gardens. *Annals of the Entomological Society of America* 101(1): 140–150.

McClintock N. and Cooper J. (2010) Cultivating the commons? An assessment of the potential for urban agriculture on public land in Oakland, California. Available at www.urbanfood.org (accessed 5 August 2017).

McKinney M.L. (2006) Urbanization as a major cause of biotic homogenization. *Biological Conservation* 127(3): 247–260.

McLain R., Poe M., Hurley P.T., Lecompte-Mastenbrook J. and Emery M.R. (2012) Producing edible landscapes in Seattle's urban forest. *Urban Forestry & Urban Greening* 11(2):187–194.

Minkoff-Zern L.A. (2012) Knowing 'good food': Immigrant knowledge and the racial politics of farmworker food insecurity. *Antipode* 46(5): 1190–1204.

Mougeot L.J. (2010) *Agropolis: The Social, Political and Environmental Dimensions of Urban Agriculture*. Routledge, London.

Pardee G.L. and Philpott S.M. (2014) Native plants are the bee's knees: Local and landscape predictors of bee richness and abundance in backyard gardens. *Urban Ecosystems* 17(3): 641–659.

Pawelek J., Frankie G.W., Thorp R.W. and Przybylski M. (2009) Modification of a community garden to attract native bee pollinators in urban San Luis Obispo, California. *Cities and the Environment* 2(1): 7–21.

Reynolds K. and Cohen N. (2016) *Beyond the Kale: Urban Agriculture and Social Justice Activism in New York City*. University of Georgia Press, Athens, GA.

Rudd H., Vala J. and Schaefer V. (2002) Importance of backyard habitat in a comprehensive biodiversity conservation strategy: A connectivity analysis of urban green spaces. *Restoration Ecology* 10(2): 368–375.

Saldivar-Tanaka L. and Krasny M.E. (2004) Culturing neighborhood open space, civic agriculture, and community development: The case of Latino community gardens in New York City. *Agriculture and Human Values* 21: 399–412.

Schwartz M.W., Thorne J.H. and Viers J.H. (2006) Biotic homogenization of the California flora in urban and urbanizing regions. *Biological Conservation* 127: 282–291.

Smit J., Nasr J. and Ratta A. (1996) *Urban Agriculture: Food Jobs and Sustainable Cities*. United Nations Development Programme (UNDP), New York.

Smith R.M., Thompson K., Hodgson J.G., Warren P.H. and Gaston K. J. (2006) Urban domestic gardens (IX): Composition and richness of the vascular plant flora, and implications for native biodiversity. *Biological Conservation* 129: 312–322.

Swan C.M., Swan C.M., Pickett S.T.A., Szlavecz K., Warren P., Willey T. (2011) Biodiversity and community composition in urban ecosystems: Coupled human, spatial and metacommunity processes. In Niemelä J., Breuste J.H., Guntenspergen G., McIntyre N.E., Elmqvist T. and James P. (eds), *Urban Ecology: Patterns, Processes, and Applications*, pp. 179–186. Oxford University Press, New York.

Taylor J.R. and Lovell S.T. (2012) Mapping public and private spaces of urban agriculture in Chicago through the analysis of high-resolution aerial images in Google Earth. *Landscape and Urban Planning* 108(1): 57–70.

Thomas B.J. (2010) Food deserts and the sociology of space: Distance to food retailers and food insecurity in an urban American neighborhood. *International Journal of Human and Social Sciences* 5(6): 400–409.

Tscharntke T., Klein A.M., Kruess A., Steffan-Dewenter I. and Thies C. (2005) Landscape perspectives on agricultural intensification and biodiversity: Ecosystem service management. *Ecology Letters* 8(8): 857–874.

Treuhaft S., Hamm M.J. and Litjens C. (2009) *Healthy Food For All: Building Equitable and Sustainable Food Systems in Detroit and Oakland*. PolicyLink, Oakland, CA.

ver Ploeg M., Breneman V., Farrigan T., Hamrich K., Hopkins D., Kaufman P., Lin B., Nord M., Smith T., Williams R., Kinnison K., Olander C., Singh A., Tuckermanty E., Kratz-Kent R., Polen C., McGowan H. and Kim, S. (2009) *Access to Affordable and*

Nutritious Food: Measuring and Understanding Food Deserts and Their Consequences. Report to USA Congress. USDA Economic Research Service.

Whittinghill L.J. and Rowe D.B. (2012) The role of green roof technology in urban agriculture. *Renewable Agriculture and Food Systems* 27(04): 314–322.

Wooten H. (2008) Food system meta-analysis for Oakland, California. *Public Health Law & Policy/Food First*, Oakland.

Zasada I. (2011) Multifunctional peri-urban agriculture: A review of societal demands and the provision of goods and services by farming. *Land Use Policy* 28(4): 639–648.

Zezza A. and Tasciotti L. (2010) Urban agriculture, poverty, and food security: Empirical evidence from a sample of developing countries. *Food Policy* 35(4): 265–273.

6

URBAN BIOLOGICAL INVASIONS

When vertebrates come to town

Lucas A. Wauters and Adriano Martinoli

Introduction

Since our species, *Homo sapiens*, changed its foraging behaviour from a migratory collector of plants and invertebrates and hunter to a sedentary farmer–hunter, it started to transform the landscape in a profound way. These land use changes have increased drastically over the past 3000 years, the so-called Anthropocene, culminating in wide-scale transformation or replacement of natural habitats with man-made ones, with industrialised and urbanised areas undergoing the most extreme changes (Baker and Harris 2007; Russo and Ancillotto 2015). A well-studied example of Early Anthropocene is the Mayacene that occurred from ca. 3000 to 1000 years BP. Late Quaternary paleo-environmental records provide evidence that Maya impacted on climate, vegetation and hydrology, altering local and regional ecosystems and landscapes, mainly through vast urban infrastructures and the creation of reservoir wetlands and canals (Beach *et al.* 2015). Deforestation was moderate, although an increasing amount of forest was transformed into agricultural land and urban settlements towards 1700–1100 BP.

Globally, the Anthropocene is also characterised by man-driven loss of biodiversity, referred to as the Planet's sixth mass extinction of species. In this era, humans caused population (species) extirpations or critical declines in local species abundance; for example 322 species of terrestrial vertebrates have gone extinct since 1500 (Dirzo *et al.* 2014). There is little doubt that such animal declines will have a cascading effect on ecosystem functioning and human well-being.

Many studies have revealed the negative or even detrimental effects of urbanisation on animal species and sometimes even on entire communities. From specific to more general effects, these include: (1) direct human interference comprising artificial illumination and noise (Francis and Barber 2013); (2) large numbers of cats and dogs, that are important predators especially for reptiles, small birds and mammals; (3) high road mortality and/or reduced capacity of movements or dispersal due to barriers (Wauters *et al.* 1997; Baker and Harris, 2007); (4) high concentrations of chemical and physical pollutants affecting animals (Perugini *et al.* 2011); (5) loss and fragmentation of suitable habitat. Nevertheless, the attentive city visitor or inhabitant will encounter different animals, some of them even in extremely high numbers; in fact urbanisation favours some generalist species that are highly successful under human-altered conditions (van Rensburg *et al.* 2009; Magle *et al.* 2012; Russo and Ancillotto 2015). What do these species have in common? Most have a wide ecological niche, thus being able to cope with, for example, a poor variety of food resources or extreme temperature conditions, and they often show a high degree of phenotypic (behavioural, physiological or even morphological) plasticity.

Not surprisingly, these are the same characteristics that affect the invasiveness of introduced alien species (Francis and Chadwick 2015; Early *et al.* 2016), hence their capacity to adapt to and thrive in the new environment. Consequently, many invasive alien species (IAS) occur more frequently and/or are more abundant (higher densities) in urban (or suburban) than in more natural habitats, as is the case for the so-called 'synurbic' native species (van Rensburg *et al.* 2009; Adams *et al.* 2013; Francis and Chadwick 2015). This combination of urbanisation and biological invasions often leads to a situation in which a small number of species that are well adapted to human-dominated landscapes, and that often are highly successful alien invaders (Clergeau *et al.* 2006; McKinney 2006), replace a wider range of native species. This pattern, termed *biotic homogenization*, is occurring in several regions across the globe and, because of the strong reduction in species richness and consequent disruption of food webs and energy flows, is known to reduce the resilience of ecosystems to environmental change (Clergeau *et al.* 2006; McKinney 2006; van Rensburg *et al.* 2009).

There is now a vast amount of evidence that invasive alien species are a global threat for biodiversity and human livelihoods, and the increasing globalisation facilitates IAS arrival, while environmental changes and climate change, facilitate IAS establishment (Early *et al.* 2016).

Why and from where are vertebrates colonising urban areas?

Urban areas tend to be colonised by species originating from many different ecosystems: such an atypical mixture of species is often referred to as 'recombinant' communities (Francis and Chadwick 2015). Many of the vertebrate species in towns are concentrated in small habitat patches (mainly parks and larger gardens). Next to processes that promote the colonisation by native vertebrates, such as habitat heterogeneity and connectivity with surrounding more natural habitats, it is the intentional or accidental introduction of alien species by man that shapes animal communities in cities. For vertebrates, one of the major pathways is the deliberate, although often illegal, importation of non-native animal pets that are subsequently (illegally) released or escape from captivity. For aquatic vertebrates also water-way infrastructures can facilitate the arrival of alien species.

Not all species thrive in urban areas: animals common in cities are those that can exploit a wide variety of resources (generalist species), or non-domesticated species closely associated with humans (synanthropes) (Francis and Chadwick 2015). Not all alien species persist in the new (urban or other) environment. Many others do survive and may even thrive, but without detrimental effects on native fauna or human health and/or activities. However, there is always a small but important proportion of alien species that do cause problems, called invasive alien species (IAS).

One group of vertebrates for which the effects of urbanisation are quite well documented are birds. Several studies have shown that bird communities tend to have three main characteristics: (1) relatively undisturbed areas just outside the city consisting mainly of native vegetation are dominated by urban avoiders, that are mostly native species; (2) in the suburban habitats (intermediate levels of urbanisation) species richness is often highest, with both native and alien species, so-called 'suburban adapters' and (3) in the most urbanised areas such as city centres, only a small number of species, many of which are IAS, persist, called the 'urban exploiters' (Blair 1996; van Rensburg et al. 2009). For example, in a study of avian communities across an urban gradient (urban, suburban and semi-natural) in Pretoria (South Africa), species richness was higher in the semi-natural than in the urban zone but the opposite was true for bird abundance (van Rensburg et al. 2009). The increase in abundance was mainly caused by alien species, with the three most abundant species in the urban zone all being aliens: the rock dove (*Columba livia*), the common myna (*Acridotheres tristis*) and the house sparrow (*Passer domesticus*).

Vertebrate species and human empathy: an 'explosive' combination

One of the worst IAS, especially in urban and suburban areas, is the domestic cat (*Felis catus*). Free-ranging cats are a global threat to a wide variety of small vertebrates, in particular in urban and suburban areas and on islands (Medina et al. 2011; Elizondo and Loss 2016). There are three types of free-ranging cats: animals not owned by and independent of humans (feral cats); cats living outdoors but partially fed and/or sheltered by humans (semi-feral cats); and cats owned by people but given outdoor access (free-ranging pet cats). In the USA, humans keep about 90 million pet cats, of which 40–80 per cent (36–72 million) are allowed outdoors, but estimates of the abundance of feral and semi-feral cat populations (30–120 million) remain purely speculative (Lepczyk et al. 2010; Elizondo and Loss 2016). Free-ranging cats cause direct wildlife mortality by predation (Baker et al. 2005; Van Heezik et al. 2010) and the numbers they kill are impressive: in the USA it is estimated that each year about 1.3–4 billion birds and 6.3–22.3 billion mammals are killed by cats (Elizondo and Loss 2016). Many free-ranging cats also kill large numbers of amphibians and reptiles, and some preliminary estimates of annual mortality in the USA are in the hundreds of millions (Elizondo and Loss 2016). But cats can also have more subtle indirect effects (*fear effects*), that alter animal behaviour and can decrease the fecundity of some prey species (Bonnington et al. 2013). These numbers of kills are impressive and it is typical for the attitude of a large portion of the public that animal welfare activists do not bother about this 'extermination' caused by one of man's most-loved animals, but are immediately alarmed and opposed to IAS management plans that plan to eliminate a few hundred or a few thousand introduced raccoons or grey squirrels (Bertolino *et al.* 2016; La Morgia *et al.* 2016).

There is no doubt that feral cat numbers must be controlled: decisions on where, when and how to control should be based on reliable information about free-ranging cat numbers and successive estimates of their predation, thus on the threat cats pose to wildlife populations. Quantifying cat abundance with rigorous and standardised methods is necessary to evaluate whether various management measures success-fully reduce population size. Unfortunately, in most cities cat management remains a controversial issue. The only method that gains at least some support by (part of) the public is a trap–neuter–return (TNR) programme: feral and semi-feral cats are captured, brought to a vet where they are sterilised (neutering), vaccinated and marked (e.g. by ear tipping), and once recovered, released at the point of capture.

Some examples on emblematic cases in the world

Grey squirrels in Europe

The grey squirrel (*Sciurus carolinensis*), native to large parts of North America, is among the world's worst IAS (Bertolino *et al.* 2008; Gurnell *et al.* 2016; La Morgia *et al.* 2016). In Europe, it is widespread in the UK, Ireland and large parts of north-ern Italy, from where it is likely to spread across much of continental Europe (Bertolino *et al.* 2008). In this non-native range it is one of the most frequently encountered mammals in urban habitats (Baker and Harris 2007; Merrick *et al.* 2016). Being attractive furry animals with conspicuous behaviour, most people in cities have a favourable attitude to the grey squirrel (64 per cent), whereas this pro-portion is much lower in rural and forested areas where it is considered a pest by the vast majority of landowners/foresters. In Europe, grey squirrels cause significant economic and social costs: damaging trees through bark stripping and raiding maize fields close to woodlands to feed on the seeds; entering buildings where they gnaw insulation and electrical wiring; binging on supplementary bird feeders removing large quantities of food. From an ecological point of view, increasing and spreading populations of grey squirrels typically cause local extinction of the native Eurasian red squirrel (*Sciurus vulgaris*) through resource competition and, in most of the UK and parts of Ireland, increased disease risk in the native species through parasite (mainly squirrel poxvirus) transmission (Tompkins *et al.* 2002; Gurnell *et al.* 2004). Some studies suggest that grey squirrels may also negatively influence bird popula-tions through competition for food, and/or nest predation (Gurnell *et al.* 2016).

A recent study carried out in Sheffield (UK) assessed which factors limit grey squirrel distribution and abundance in urban environments (Bonnington *et al.* 2014). The total city-wide population was estimated at 6539 adult squirrels (0.46 squirrels/ha), but in good-quality habitats, maximum densities reached more than eight animals/ha. These sites were characterised by denser canopy cover, availability of seed bearing trees and supplementary feeders, provided for garden birds. Dense urbanised areas with little green were avoided. Thus, grey squirrels in towns forage in gardens and parks, areas known to support the highest biodiversity in urbanised areas, and thereby give city dwellers some kind of 'connection' with nature which explains the positive feeling that is created between grey squirrels and man in most cities in its non-native range.

Is efficient grey squirrel control/eradication possible in urban areas? Although with the new EU regulation this is now a legal obligation in Italy, eradication programmes of this charismatic mammal generally encounter strong opposition (e.g. Bertolino *et al.* 2016). For the management of the central-Italian population in and around Perugia, La Morgia *et al.* (2016) developed a structured decision-making technique based on a Bayesian decision network model that explicitly considered the diversity in public attitude to reduce expected social conflicts. Where the model identified priority areas where rapid eradication is fundamental, expanding it to integrate the attitudes of citizens towards the project resulted in toning down the main strategy, causing a reduction of the overall utility of interventions (La Morgia *et al.* 2016). Only the future will tell us whether this approach, which combines the evaluation of technical efficiency, feasibility and social impact on control management decisions, will allow the goals set by the new EU regulation to be reached.

Rats and mice: past and current problems

For thousands of years, the most widely spread commensal rodents such as the black rat (*Rattus rattus*), the Norway rat (*Rattus norvegicus*), the house mouse (*Mus domesticus*) and the pacific rat (*Rattus exulans*), have lived in close association with humans. As human settlements expanded and became urbanised, several of these rodents spread with humans to new locations around the globe. Rats can become predators of bird eggs and nestlings, and both rats and house mice can devastate stocked crops, especially wheat and maize (Banks and Smith 2015). However, the worst risks of these IAS are human health risks. In fact, rats are a source for some very dangerous zoonotic pathogens responsible for significant human morbidity and mortality, and rat-associated health risks are particularly problematic in urbanised areas (Himsworth *et al.* 2013). Urban Norway and black rats can host zoonotic bacteria (*Leptospira interrogans*, *Yersina pestis*, *Rickettsia typhi*, *Bartonella* spp., *Streptobacillus moniliformis*), viruses (Seoul hantavirus) and macroparasites (*Angiostrongylus cantonensis*). Given the current rates of global urbanisation, it has become a priority to develop a thorough and modern understanding of rat-associated zoonoses in urban centres to determine human health risks and develop strategies to both monitor and mitigate those risks (Himsworth *et al.* 2013).

Efficient monitoring depends in a large part on proper awareness among healthcare professionals regarding the occurrence of and risk-factors associated with rat-associated diseases to avoid misdiagnoses. Finally, effective rat control strategies and other public health measures need to be carried out to prevent or at least reduce rat-to-human disease transmission (Himsworth *et al.* 2013).

Animal pet trade: a dangerous business

Several well-known examples of IAS are exclusively or mainly linked with the pet trade and subsequent escape from captivity or deliberate (in most countries illegal) release of these species. Here we illustrate some well-known examples.

The invader does not always win

A case-study of the interactions between native (*Gehyra australis*, Gekkonidae) and invasive alien gecko lizards (*Hemidactylus frenatus*, Gekkonidae) in urban environments of tropical Australia shows that the invasive species do not always win (Yang *et al.* 2012). Geckos are small commensal animals that often find their way into cargos and thus can be transported over long distances; consequently several species of the genus *Hemidactylus* have been introduced to a wide range of areas in Asia, Africa, North and South America, and Australasia (Yang *et al.* 2012). Furthermore, geckos, as many other reptiles, occupy a big part of the animal pet-trade market. These invaders have caused marked declines in the abundance and distribution of native gecko species in many countries. However, in a rural village near Darwin, Australia, the alien *Hemidactylus frenatus* is sympatric with the larger, native *Gehyra australis;* both are abundant and forage on the same buildings. Laboratory encounters showed that during encounters the alien species is subordinate to the larger native species. There is also niche partitioning in time: *H. frenatus* is mostly active during the dry-season, while *G. australis* prefers the wet-season (Yang *et al.* 2012). Where the two taxa co-occurred, invasive *H. frenatus* modified substrate use in the presence of the native *G. australis*, consistent with competitive displacement. In this case, the invasive gecko appears to be exploiting a 'vacant niche' around buildings, rather than displacing the native gecko taxon. However, this particular outcome of interspecific competition may be due to the larger body size of the native species and the invader might have a significant impact on smaller native lizards (Yang *et al.* 2012).

Subtle effects, difficult to discover

Parrots (order Psittaciformes) are among the most traded and introduced birds worldwide. Many species nest in trunk cavities potentially competing with native hole-breeding birds, arboreal rodents and bats (Menchetti and Mori 2014; Appelt *et al.* 2016). Parrots and parakeets are known to damage crops and electrical infrastructures and are potential reservoirs of *Chlamydophila psittaci*, the etiological agent of human psittacosis, and other diseases transmittable to humans and wildlife. Different studies are tackling the problem of when and where control/eradication is useful or necessary (e.g. Strubbe *et al.* 2010) but in several cases where management was deemed necessary, eradication programmes were often hampered by emotional reactions of part of the public.

One of the most common introduced parakeet species is the ring-necked parakeet (*Psittacula krameri*) native in Africa and Asia, which has been introduced in several European countries where it competes for nest-sites with native hole-nesting birds. In a study in Belgium, Strubbe and Matthysen (2009) showed experimentally that a native hole-nesting passerine, the nuthatch (*Sitta europaea*) declined significantly, largely due to nest take-overs by parakeets. They showed that there was a strong overlap in preferred nest sites which explained the interspecific competition that led to a reduction in nuthatch numbers when suitable cavities became scarce (Strubbe and Matthysen 2009). Their study raised serious concern for the conservation of native cavity-nesting birds when invasive parakeet populations are allowed to expand further their range. However, using species distribution models combined with empirical estimates of competition strength between this IAS and native nuthatches, Strubbe *et al.* (2010) concluded that there is no compelling evidence that parakeets pose a large enough threat to justify an eradication campaign in the areas in Belgium currently occupied. In contrast, studies of urban population of ring-necked parakeets in Spain strongly suggest this IAS negatively affects some rare, already (locally) endangered, species due to competition for suitable nest sites; a bat, the greater noctule (*Nyctalus lasiopterus*) which roosts in tree cavities and a raptor bird, the lesser kestrel (*Falco naumanni*), a colony breeder using cavities in walls for breeding (Hernàndez-Brito *et al.* 2014).

Urban species and their critical threats

IAS can quickly spread and affect nearly all terrestrial and aquatic ecosystems. They have become one of the greatest environmental challenges of the twenty-first century with extreme consequences also in economics and human health costs, with an estimated total cost (including their direct control/eradication) of more than $120 billion per year in the USA (www.fort.usgs.gov/branch/100, accessed on 27.12.2016). In Europe, recent estimates of direct costs due to IAS reach at least €12.7 billion per year (Kettunen *et al.* 2009).

Critical threats from urban IAS area damage to structures, impact on human activities and human health, and loss of urban biodiversity.

A pest species for many reasons: the example of the house crow Corvus splendens

House crows are a notorious pest both within the native range, the Indian subcontinent and in countries where it has been introduced (e.g. Malaysia, Singapore, Zanzibar, Mauritius and Arabian Peninsula). The species is particularly successful in invading urban landscapes, less so in colonising rural landscapes (Wilson *et al.* 2015). House crows are opportunistic omnivores and scavengers, and as such they can have an important ecological role consuming carcasses that otherwise may attract flies and parasites. However, in their native range they heavily damage agriculture crops (in particular sunflowers and maize; Wilson *et al.* 2015) causing huge economic losses. In countries where house crows are an IAS, they can damage agriculture, predate native species, compete with native corvids, pose a threat to human health and become a nuisance to humans in urbanised areas. Being a host to the fungus *Cryptococcus neoformans* (which can be dangerous for debilitated people, infecting the lungs or the central nervous system) and *Salmonella* sp., it also plays an important role as a human disease vector.

Invasive alien species threatening human health

IAS can also become a serious problem for human health and/or for the health of domestic animals. Typical examples are the raccoon (*Procyon lotor*) (Beltrán-Beck *et al.* 2012) and the raccoon dog (*Nyctereutes procyonoides*) introduced in some European countries. Both species often settle near or in urbanised areas where they may come into direct contact with man and domestic animals, in particular dogs. The raccoon dog is a canid indigenous in East Asia and introduced in Europe more than half a century ago. In a study of its parasite fauna carried out in Estonia (255 carcasses examined), a total of 17 helminth species were recorded (4 trematodes, 4 cestodes and 9 nematodes; Laurimaa *et al.* 2016). Average parasite species richness was 2.86 with a maximum of nine species in a single host. The most prevalent parasite species were *Uncinaria stenocephala* (97.6 per cent) and *Alaria alata* (68.3 per cent).

Zoonotic hookworms (*Uncinaria*) can be transmitted to dogs and cats and also to humans when larvae penetrate unprotected skin. This can result in a disease called cutaneous larva migrans (CLM), when the larvae migrate through the skin and cause inflammation. *Alaria* species can infect for example pigs, making the meat no longer safe for human consumption. When eating infected, undercooked meat this parasite can cause severe allergic reactions in humans, and even death. However, a wider-scale screening of this IAS in Europe reports as many as 32 helminth species (mainly nematodes or trematodes), 19 of which can be considered zoonotic. Because of this high number of zoonotic parasites in an IAS that is increasing its range and population size, the raccoon dog should be considered an important source of environmental contamination with zoonotic agents in Europe (Laurimaa *et al.* 2016).

Raccoons are opportunistic carnivores native to North America that have been introduced and escaped as pets in many European countries with established populations from Spain and North Italy, through most of Central Europe up to Russia. In their native range raccoons host a number of disease agents that can be transmitted to humans, domestic animals and other wildlife (Beltrán-Beck *et al.* 2012). In several countries in NE Europe, raccoons are a reservoir of the *Lyssavirus* which causes rabies, a progressive fatal and incurable zoonotic viral encephalitis, with a total of 142 rabies cases reported in raccoons during 1990–2010 (Beltrán-Beck *et al.* 2012). Another viral disease is the canine distemper virus (CDV), which can cause an often lethal disease among wild and domestic carnivores. CDV can be epidemic among raccoons and subsequently spread to other wildlife. Raccoons have also been shown to carry leptospires in their kidneys, and serum antibodies have been detected in several countries. Hence, since they can act as a reservoir of infection by Leptospirosis bacteria, they become a risk for humans that come into close contact with them.

Probably the most studied reservoir role of raccoons is that of the raccoon round-worm (*Baylisascaris procyonis*): infected raccoons can shed millions of eggs daily, leading to widespread and heavy environmental contamination (Gavin *et al.* 2005). Prevalence of roundworm infection among introduced raccoons can vary greatly between countries: it is high in Germany, but not detected in Japan (Beltrán-Beck *et al.* 2012). *B. procyonis* is considered the most common and widespread cause of clinical larva migrans in animals, usually associated with fatal or severe neurological disease (Gavin *et al.* 2005). In humans, the raccoon roundworm is a rare cause of ocular and visceral larva migrans infection, and can also cause a fatal cerebrospinal infection in infants and young children (Gavin *et al.* 2005; Beltrán-Beck *et al.* 2012). Hence the fast expansion of raccoons in Europe risks becoming a serious threat to human health due to its role as a host to many dangerous viral, bacterial and helminth-caused diseases.

An important zoonotic disease in many parts of Europe is Lyme disease or Lyme Borreliosis. There are numerous vertebrate reservoirs for *Borrelia burgdorferi* sensu lato (sl), the etiological agents of Lyme Borreliosis (LB); one of these is the Siberian chipmunk (*Tamias sibiricus*), a rodent originating from Asia and intention-ally released into the wild in Europe since the 1970s (Vourch *et al.* 2007). In a suburban forest near Paris, France, this alien species was heavily infected with both larvae and nymphs of the LB vector, the hard tick (*Ixodes ricinus*). In this area, all chipmunks were infested with larvae and larval abundance was much higher than in the two native rodents (bank vole and wood mouse). Nymph prevalence in chipmunks was 86 per cent and the infection prevalence of *B. burgdorferi* sl was 33 per cent, much higher than in the native rodents, suggesting that the Siberian chipmunk may be an important reservoir host for LB (Vourch *et al.* 2007).

How to contrast the problem

From an ecological point of view, IAS can affect native species through predation, competition and disease transmission, but also indirectly through processes such as habitat disruption. The best course of action is preventing introductions (in par-ticular deliberate ones) by strict laws and rules on animal (pet) trade and detention of animal species in captivity (e.g. Rabitsch *et al.* 2016). Prevention avoids costly and difficult control and eradication efforts; however, once an IAS is established, identifying its effects is essential to take decisions about whether or not such efforts are prudent or necessary. Since there is a multitude of invasive species that cur-rently threaten native species and ecosystems, IAS distribution models combined with estimates of their impact (based on evidence from empirical studies) should be used as a general tool to make an assessment of the potential impact of established invasive species. Such information is required to make effective decisions on how to prioritise management effort and resources (Strubbe *et al.* 2010).

This brings us to the concept of invasive alien species indicators that provide vital information on the status quo and trends of biological invasions and on the efficacy of response measures, required by the policy sector involved in biodiversity management and conservation (Rabitsch *et al.* 2016). Unfortunately, for many alien species, these indicators lack crucial data availability and/or quality.

Since 2014, the EU has had specific legislation to contrast urban and non-urban IAS, which has a need for tools not only to monitor its efficacy but also to suggest future improvements (Beninde *et al.* 2015).

This legislation provides a legal basis for dealing with IAS; in addition, for prevention, eradication, management and control it involves scientists, stake-holders and the public in decision making processes. An important aspect is the 'polluter-pays' principle: costs of damage induced by IAS (and costs for restoration) should be supported by those who are responsible for the introduction of the species (Beninde *et al.* 2015). These can be single persons (traders), companies, local administrations or even countries.

Rabitsch *et al.* (2016) presented a set of six IAS indicators for Europe: (1) an index of invasion trends, (2) an indicator on pathways of invasions, (3) the Red List Index of IAS, (4) an indicator of IAS impacts on ecosystem services, (5) trends in incidence of livestock diseases and (6) an indicator on the costs associated with alien species management and research. They admit that each single indicator has particular strengths but also shortcomings; however, when combined they should allow for a detailed understanding of the status and trends of biological invasions in Europe. At its current stage, the IAS indicator is only a starting point, and its application on the continental scale has revealed the need for better data collection and management, and interdisciplinary collaboration (scientists, public and private stakeholders, policy sector, administrations at different levels) to improve the interoperability among existing databases and between data holders (Rabitsch *et al.* 2016). A somewhat different approach has been proposed by scientists and wildlife/conservation managers in the Czech Republic, where priority-based lists of prominent IAS that attain a high level of concern and are subject to priority monitoring and management have been produced. These so-called Black, Grey and Watch (alert) lists represent a starting point for setting priorities in prevention, early warning and management systems (Pergl *et al.* 2016).

To contrast these problems related to IAS fast and efficiently, each country should have (or create) an Invasive Species Science Team or Centre (for example the Invasive Species Science (ISS) Branch of the Fort Collins Science Center, USA). Such centres should carry out research and provide the technical assistance related to management actions of IAS. Basic background information includes definition of the pathways through which these species are introduced; identifying the areas vulnerable to invasion; forecasting invasion dynamics, constructing models to understand and predict invasive species distributions, and, as a final step, developing eradication or control methods for direct management of the IAS (e.g. Genovesi 2005; Beninde *et al.* 2015). Other tasks of such IAS centres will have to include the development of communication platforms (web sites, DVDs on case studies, press contacts), that guarantee the dissemination and sharing of information with different stakeholders (administrations responsible for wildlife control/management, other agency partners – such as public health and zootechnical agencies, municipalities, environmental and animal welfare organisations, wildlife/nature and human health oriented ONGs, hunters and farmers, and the general public).

In fact, a study carried out in the Donana social–ecological system in South West Spain demonstrated that understanding the human dimension of invasions is critical to effectively tackle the problems associated with invasive alien species (Garcia-Llorente *et al.* 2008). Using questionnaires to evaluate the social perceptions and attitudes of different stakeholder groups affected by invasive alien species they identified five stakeholder groups that greatly differed in their degree of knowledge, perceptions, attitudes, and willingness to pay for eradication. The authors concluded that such remarkably different perceptions about the impacts of IAS and attitudes toward their introduction or eradication should be considered in any decision-making process regarding their management, and consultation with different stakeholders should be encouraged from the start (Garcia-Llorente *et al.* 2008). The latter will be extremely important in the cases of highly visible and 'cute' invasive mammals and birds in urban areas where the overall approach to these species by the general public, despite the many campaigns in favour of correct and scientifically based information about the problems they create, remains one of sympathy, or even attraction, if not one of tolerance combined with indifference to the fate of native species suffering from the alien's presence. Developing appropriate information and education programmes is essential to facilitate successful implementation of management actions.

References

Adams A.L., Dickinson K.J.M., Robertson B.C. and van Heezik Y. (2013) Predicting summer site occupancy for an invasive species, the Common Brushtail Possum (*Trichosurus vulpecula*), in an urban environment. *PLoS ONE* 8(3): e58422.

Appelt C.W., Ward L.C., Bender C., Fasenella J., Van Vossen B.J. and Knight L. (2016) Examining potential relationships between exotic monk parakeets (*Myiopsitta monachus*) and avian communities in an urban environment. *The Wilson Journal of Ornithology* 128: 556–566.

Baker P.J., Bentley A.J., Ansell R.J. and Harris S. (2005). Impact of predation by domestic cats *Felis catus* in an urban area. *Mammal Review* 35: 302–312.

Baker P.J. and Harris S. (2007) Urban mammals: What does the future hold? An analysis of the factors affecting patterns of use of residential gardens in Great Britain. *Mammal Review* 37: 297–315.

Banks P.B. and Smith H.A. (2015) The ecological impacts of commensal species: Black rats, *Rattus rattus*, at the urban–bushland interface. *Wildlife Research* 42: 86–97.

Beach T., Luzzadder-Beach S., Cook D., Dunning N., Kennett D.J., Krause S., Terry R., Trein D. and Valdez F. (2015) Ancient Maya impacts on the Earth's surface: an Early Anthropocene analog? *Quaternary Science Reviews* 124: 1–30.

Beltrán-Beck B., García F.J. and Gortázar C. (2012) Raccoons in Europe: disease hazards due to establishment of an invasive species. *European Journal of Wildlife Research* 58: 5–15.

Beninde J., Fischer M.L., Hochkirch A. and Zink A. (2015) Ambitious advances of the European Union in the legislation of Invasive Alien Species. *Conservation Letters* 8: 199–205.

Bertolino S., Lurz P.W.W., Sanderson R. and Rushton S.P. (2008) Predicting the spread of the American grey squirrel (*Sciurus carolinensis*) in Europe: A call for a co-ordinated European approach. *Biological Conservation* 141: 2564–2575.

Bertolino S., Lurz P.W.W., Shuttleworth C.M., Martinoli A. and Wauters L.A. (2016) The management of grey squirrel populations in Europe: Evolving best practice. In Shuttleworth C.M., Lurz P.W.W. and Gurnell J. (eds), *The Grey Squirrel: Ecology & Management of an Invasive Species in Europe*, pp. 495–514. European Squirrel Initiative, UK.

Blair R. (1996) Land use and avian species diversity along an urban gradient. *Ecological Applications* 6: 506–519.

Bonnington C., Gaston K.J. and Evans K.L. (2013) Fearing the feline: Domestic cats reduce avian fecundity through trait-mediated indirect effects that increase nest predation by other species. *Journal of Applied Ecology* 50: 15–24.

Bonnington C., Gaston K.J. and Evans K.L. (2014) Squirrels in suburbia: Influence of urbanisation on the occurrence and distribution of a common exotic mammal. *Urban Ecosystems* 17: 533–546.

Clergeau P., Jokimäki J. and Snep R. (2006) Using hierarchical levels for urban ecology. *Trends in Ecology and Evolution* 21: 660–661.

Dirzo R., Young H.S., Galetti M., Ceballos G., Isaac N.J.B. and Collen B. (2014) Defaunation in the Anthropocene. *Science* 345: 401–406.

Early R., Bradley B.A., Dukes J.S., Lawler J.J., Olden J.D., Blumenthal D.M., Gonzalez P., Grosholz E.D., Ibanez I., Miller L.P., Sorte C.J.B. and Tatem A.J. (2016) Global threats from invasive alien species in the twenty-first century and national response capacities. *Nature Communications* 7: 12485.

Elizondo E.C. and Loss S.R. (2016) Using trail cameras to estimate free-ranging domestic cat abundance in urban areas. *Wildlife Biology* 22: 246–252.

Francis C.D. and Barber J.R. (2013) A framework for understanding noise impacts on wildlife: An urgent conservation priority. *Frontiers in Ecology and the Environment* 11: 305–313.

Francis R.A. and Chadwick M.A. (2015) Urban invasions: Non-native and invasive species in cities. *Geography* 100: 144–151.

Garcia-Llorente M., Martin-Lopez B., Gonzales J.A., Alcorlo P., Montes C. (2008) Social perceptions of the impacts and benefits of invasive alien species: Implications for management. *Biological Conservation* 141: 2969–2983.

Gavin P.J., Kazacos K.R. and Shulman S.T. (2005) Baylisascariasis. *Clinical Microbiology Reviews* 18: 703–718.

Genovesi P. (2005) A strategy to prevent and mitigate the impacts posed by invasive alien species in Europe. In Nentwig W., Bacher S., Cock M.J.W., Dietz H., Gigon A. and Wittenberg R. (eds), *Biological Invasions: From Ecology to Control. Neobiota*, 6: 145–147.

Gurnell J., Wauters L.A., Lurz P.W.W. and Tosi G. (2004). Alien species and interspecific competition: Effects of introduced eastern grey squirrels on red squirrel population dynamics. *Journal of Animal Ecology* 73: 26–35.

Gurnell J., Lurz P.W.W. and Shuttleworth C.M. (2016) Ecosystem impacts of an alien invader in Europe, the grey squirrel Sciurus carolinensis. In Shuttleworth C.M., Lurz P.W.W. and Gurnell J. (eds), *The Grey Squirrel: Ecology & Management of an Invasive Species in Europe*. European Squirrel Initiative, UK, pp. 307–328.

Hernàndez-Brito D., Carrete M., Popa-Lisseanu A.G., Ibànez C. and Tella J.L. (2014) Crowding in the city: Losing and winning competitors of an invasive bird. *PLoS ONE* 9(6): e100593.

Himsworth C.G., Parsons K.L., Jardine C. and Patrick D.M. (2013) Rats, cities, people, and pathogens: A systematic review and narrative synthesis of literature regarding the ecology of rat-associated zoonoses in urban centers. *Vector-Borne and Zoonotic Diseases* 13: 349–359.

Kettunen M., Genovesi P., Gollasch S., Pagad S., Starfinger U., ten Brink P. and Shine C. (2009) *Technical Support to EU Strategy on Invasive Alien Species (IAS) – Assessment of the impacts of IAS in Europe and the EU.* Institute for European Environmental Policy, London and Brussels.

La Morgia V., Paoloni D. and Genovesi P. (2016) Eradicating grey squirrel *Sciurus carolinensis* from urban areas: An innovative decision making approach based on lessons learnt in Italy. *Pest Management Science* 73: 354–363.

Laurimaa L., Süld K., Davison J., Moks E., Valdmann H. and Saarma U. (2016) Alien species and their zoonotic parasites in native and introduced ranges: The raccoon dog example. *Veterinary Parasitology* 219: 24–33.

Lepczyk C.A., Dauphinè N., Bird D.M., Conant S., Cooper R.J., Duffy D.C., Hatley P.J., Marra E.S., Stone E. and Temple S.A. (2010) What conservation biologists can do to counter trap-neuter-return: Response to Longcore *et al.*. *Conservation Biology* 24: 627–629.

Magle S.B., Hunt V.M., Vernon M. and Crooks K.R. (2012) Urban wildlife research: Past, present, and future. *Biological Conservation* 155: 23–32.

McKinney M.L. (2006) Urbanization as a major cause of biotic homogenization. *Biological Conservation* 127: 247–260.

Medina F.M., Bonnaud E., Vidal E., Tershy B.M., Zavaleta E.S., Donlan C.J., Keitt B.S., Le Corre M., Horwath S.V. and Nogales M. (2011) A global review of the impacts of invasive cats on island endangered vertebrates. *Global Change Biology* 41: 3503–3510.

Menchetti M. and Mori E. (2014) Worldwide impact of alien parrots (Aves Psittaciformes) on native biodiversity and environment: A review. *Ethology, Ecology & Evolution* 26: 172–194.

Merrick M.J., Evans K.L. and Bertolino S. (2016) Urban grey squirrel ecology, associated impacts, and management challenges. In Shuttleworth C.M., Lurz P.W.W. and Gurnell J. (eds), The Grey Squirrel: Ecology & Management of an Invasive Species in Europe. European Squirrel Initiative, UK, pp. 57–78.

Pergl J., Sádlo J., Petrusek A., Laštůvka Z., Musil J., Perglová I., Šanda R., Šefrová H., Šíma J., Vohralík V. and Pyšek P. (2016) Black, Grey and Watch Lists of alien species in the Czech Republic based on environmental impacts and management strategy. *NeoBiota* 28: 1–37.

Perugini M., Manera M., Grotta L., Abete M.C., Tarasco R. and Amorena M. (2011) Heavy metal (Hg, Cr, Cd, and Pb) contamination in urban areas and wildlife reserves: Honeybees as bioindicators. *Biological Trace Element Research* 140: 170–176.

Rabitsch W., Genovesi P., Scalera R., Biała K., Josefsson M. and Essla F. (2016) Developing and testing alien species indicators for Europe. *Journal for Nature Conservation* 29: 89–96.

Russo D. and Ancillotto L. (2015) Sensitivity of bats to urbanization: A review. *Mammalian Biology* 80: 205–212.

Strubbe D. and Matthysen E. (2009) Experimental evidence for nest-site competition between invasive ring-necked parakeets (*Psittacula krameri*) and native nuthatches (*Sitta europaea*). *Biological Conservation* 142: 1588–1594.

Strubbe D., Matthysen E. and Graham C.H. (2010) Assessing the potential impact of invasive ring-necked parakeets *Psittacula krameri* on native nuthatches *Sitta europaea* in Belgium. *Journal of Applied Ecology* 47: 549–557.

Tompkins D.M., Sainsbury A.W., Nettleton P., Buxton D. and Gurnell J. (2002) Parapoxvirus causes a deleterious disease in red squirrels associated with UK population declines. *Proceedings of the Royal Society London B* 269: 529–533.

Van Heezik Y., Smyth A., Adams A. and Gordon J. (2010) Do domestic cats impose an unsustainable harvest on urban bird populations? *Biological Conservation* 143: 121–130.

van Rensburg B.J., Peacock D.S. and Robertson M.P. (2009) Biotic homogenization and alien bird species along an urban gradient in South Africa. *Landscape and Urban Planning* 92: 233–241

Vourch G., Marmet J., Chassagne M., Bord S. and Chapuis J.-L. (2007) *Borrelia burgdorferi Sensu Lato* in Siberian Chipmunks (*Tamias sibiricus*) introduced in suburban forests in France. *Vector-Borne and Zoonotic Diseases* 7: 637–641.

Wauters L.A., Somers L. and Dhondt A.A. (1997) Settlement behaviour and population dynamics of reintroduced red squirrels *Sciurus vulgaris* in a park in Antwerp, Belgium. *Biological Conservation* 82: 101–107.

Wilson R.F., Sarim D. and Rahman S. (2015) Factors influencing the distribution of the invasive house crow (*Corvus splendens*) in rural and urban landscapes. *Urban Ecosystems* 18: 1389–1400.

Yang D., Gonzàlez-Bernal E., Greenlees M. and Shine R. (2012) Interactions between native and invasive gecko lizards in tropical Australia. *Austral Ecology* 37: 592–599.

7

MANAGEMENT OF PLANT DIVERSITY IN URBAN GREEN SPACES

Myla F.J. Aronson, Max R. Piana, J. Scott MacIvor and Clara C. Pregitzer

Introduction

Although cities are often considered to have a negative impact on native plant diversity, recent global and continental analyses have shown that cities support surprisingly diverse floras. Aronson *et al.* (2014) found that 112 cities supported 5 percent of the global flora, with species richness in cities dominated by native species. Threatened and endangered plant species are also found in cities (Aronson *et al.* 2014; Ives *et al.* 2016). Not surprisingly, however, cities on average retain only 25 percent of estimated pre-urban species diversity (Aronson *et al.* 2014). Planning, restoration, and management of plant communities in cities is critical to conserving and enhancing biodiversity and the ecosystem services vegetation provides in cities.

The composition and condition of urban vegetation drives trophic interactions, dictating what animal species will succeed in city environments. Recent work has found that the structure and composition of urban vegetation drives bird, bat, and insect communities, and that native vegetation supports higher levels of biodiversity than non-native vegetation (Ballard *et al.* 2015; Threlfall *et al.* 2017). The management of urban green spaces (UGS) at the city, neighborhood, and local scales affects their capacity to support biodiversity (Beninde *et al.* 2015) as well as their provision of critical ecosystem services.

Urban vegetation plays a key role in the provision of ecosystem services such as carbon sequestration, nutrient cycling, air and water purification, noise reduction, runoff mitigation, food supply, and temperature regulation (Gómez-Baggethun *et al.* 2013; Pincetl *et al.* 2013; Ahern *et al.* 2014). Urban vegetation also has recreation and cultural values (Lubbe *et al.* 2010; Ahern *et al.* 2014). For example, UGS with higher plant species richness have been positively linked to higher psychological benefits for urban users (Fuller *et al.* 2007).

TABLE 7.1 Management intensity, economic input, stakeholders, management recommendations, and ecological/human use trade-offs of common urban green spaces (UGS) as components of plant conservation in cities. Modified from Aronson et al. (2017), *Frontiers in Ecology and the Environment*.

Management intensity	Economic input	Urban green space	Stakeholders	Management for biodiversity	Limitation as habitat	Limitations for human use
low	low	Remnant habitat	Public: state, city	• Manage and design to minimize human impacts • Provide/improve connectivity to existing remnants and new habitat • Manage invasive plants	• Remnant size, edge effects, and fragmentation limit urban-sensitive species • Human use, degradation	• Not suitable for some recreational activities • Safety, perceptions of crime • May lack local stewardship
		Riparian corridors	Public: city, state	• Maintain vegetation cover, and create ecotones from aquatic to terrestrial habitats • Manage invasive plants	• Constricted/channelized rivers prevent natural function • Narrow zone of terrestrial habitat creates edge effects and prevents development of natural ecotones • Pollution • Dominance of invasive plant species (eg Himalayan balsam, Japanese knotweed) • Human use, degradation	• Access • "Messy perception" • Lack of awareness of river system • May lack local stewardship
		Abandoned railway corridors	Private: corporate Public: city, state	• Manage invasive plants • Restoration to enhance use as corridors • Increase native species diversity and vegetation structure	• Size, linear nature • Corridor for invasions • Human use	• Access restricted • Safety • Lacks local stewardship

Brownfields/ industrial parks	Private: corporate	• Remediate highly toxic contaminants • Minimize impervious surfaces • Increase the proportion and diversity of native plants • Soil remediation	• Contamination and historical use • Pavement and impervious infrastructure	• Few people can experience • Loss of cultural value • Safety • Lacks local stewardship
Vacant lots	Multiple – city, private	• Restoration with goals to increase the proportion/ diversity of native plants • Soil remediation	• Small, fragmented • Multiple uses • Ephemeral	• Safety • Ownership • May lack local stewardship
Cemetery/ graveyards	Private, city	• Minimize pesticides and chemical applications • Reduce frequency of mowing where acceptable, and elsewhere replace with short swards of native flowering plants • Restore vegetation structure	• Depends on management regime • For heavily managed sites: maintenance of turfgrass lawn	• Public perception limits range of acceptable activities
Roadside	City	• Reduce salt and pesticide applications • Increase the proportion/ diversity of native plants	• Size, linear nature, noise, salt and herbicide, air pollution	• Safety • High noise and pollution levels • Active maintenance • May lack local stewardship

(continued)

TABLE 7.1 (continued)

Management intensity	Economic input	Urban green space	Stakeholders	Management for biodiversity	Limitation as habitat	Limitations for human use
		Managed parks	City	• Encourage public participation in maintenance • Restore vegetation structure, including wetland habitats • Manage invasive plants • Reduce soil compaction and pollution	• Recreational use and municipal maintenance of turfgrass lawn	• "Messy perception" • Lack of management • Safety, perceptions of crime • Social inequities in proximity to managed parks
		Community/ allotment gardens	Multiple – city, private, public	• Plant non-cultivated flowers to provide food in all seasons for pollinators • Reduce pesticide and herbicide application	• Human activity and disturbance to soil • Ordinances restricting the planting of non-crop plants	• Area available for food production • Reduce yield due to non-invasive pest management • Availability of plots/ waiting lists
		Lawn	Multiple – homeowner, corporate, city	• Integrate more diverse plantings or replace lawn with other habitat types with higher biodiversity value • Reduce mowing frequency/create a mosaic of amenity grassland of different heights • Reduce pesticide and herbicide application	• Monoculture of few grass species • Frequent trampling • Designed exclusively for human use, aesthetics • Impacts of pesticide and herbicide use	• Aesthetic impact on some citizens • Dry season, watering • Human health (pesticides) • Limited access when privately owned (corporate campuses, private homeowners)

Home garden	Homeowner	• Reduce lawn area • Natives and non-invasive ornamentals	• Often fragments • Only desirable species welcome • Not managed at city scale	• Private • Few users
Green roof	City, private business	• Vary substrate, plant community • Include more native species adapted to conditions. • Provide cover; logs, rocks, features to protect against wind and sun • Mimic native soils, mycorrhizae	• Height, small size, proximity to urban green space • Accessible only to certain species	• Increasing cost, maintenance • Access restricted • May lack local stewardship
Bioswale	City	• Reduce litter accumulation • Maintain drainage • Increase the proportion/diversity of native plants • Mimic native soils, mycorrhizae	• Small, fragmented	• Not designed for human use • May lack local stewardship
Green wall	City, business	• Full cover; surface area • 3D structure to increase complexity for protection	• Size of vegetated area, minimal or no substrate	• Increasing cost, maintenance • Access restricted • May lack local stewardship

high →

high →

The ability of cities to support both plant and animal biodiversity as well as eco-system services is contingent on the amount and quality of UGS (Cilliers *et al.* 2013, Aronson *et al.* 2014, Beninde *et al.* 2015). Urban green spaces are defined as all natural, semi-natural, and artificial ecological systems within a city (Cilliers *et al.* 2013) and range from native remnant habitats to brownfields, vacant lots, gardens and yards, to engineered green infrastructure including bioswales and green roofs (Aronson *et al.* 2017). Management of vegetation in UGS ranges in intensity, and decisions made for biodiversity conservation and human use often conflict (Table 7.1). However, most UGS, from privately owned home gardens to large public parks, can be designed and managed to support both people and biodiversity.

What influences plant community composition and structure in cities?

Understanding the factors that influence floras in cities is imperative to developing effective management plans to sustain plant diversity and the ecosystem services plants provide. Urban floras face a series of stressors that can be divided into five main categories: habitat transformation, habitat fragmentation, the urban environment, non-native species introductions, and management decisions (Williams *et al.* 2009). The stressors on urban floras are more or less important depending on the type of UGS and taxa-specific life history requirements. Habitat transformation—from natural to urban habitats (e.g., built structures, vacant lots, brownfields, lawns) results in the loss of many species (Williams *et al.* 2009). Particular habitats are more susceptible to loss or transformation, including flatlands and wetlands (Williams *et al.* 2009). For example, the Cape Flats Sand Fynbos is a critically endangered vegetation type that lies entirely within the municipal boundaries of Cape Town, South Africa (Rebelo *et al.* 2011). This vegetation type is only found on areas with flat topography that have undergone rapid urbanization since the 1960s, resulting in an 85 percent loss of this vegetation type and only 1.5 percent of the remaining habitat is managed for conservation. On the other hand, the Peninsula Sandstone Fynbos vegetation type, also located entirely within the municipal boundaries of Cape Town, is well conserved, due to its location on rocky, steep slopes and variable topography, rendering this habitat undevelopable (Rebelo *et al.* 2011). Only 2.5 percent of the Peninsula Sandstone Fynbos has been transformed, and close to 80 percent is managed for conservation. Wetlands are another habitat preferentially lost to urbanization (Ehrenfeld 2000; Gibbs 2000). Between 1992 and 2012, urban expansion in China let to a greater loss of wetlands than any other natural land cover (He *et al.* 2014).

Habitat fragmentation decreases connectivity and dispersal dynamics, altering the type and number of spontaneous plant colonizers that can contribute positively or negatively to the management of urban plant communities. In the New York City (NYC) metropolitan area, increases in human population density along urban–rural gradients were related to decreases in the number and area of individual wetlands, and increases in the distance between them (Gibbs 2000). The implications of smaller wetlands and increased isolation are a lack of dispersal among wetlands, effectively disconnecting populations and hindering metapopulation dynamics. Some plant species may be more susceptible to local extinction when metapopulations are disrupted by fragmentation, particularly those with low dispersal ability and obligate pollinator, seed dispersal, or mycorrhizae mutualisms (Ghazoul 2005; Williams et al. 2009). Even plant species with generalist pollinators may experience pollen limitation and reduced seed set due to fragmentation of urban environments (Hennig and Ghazoul 2011). For example, small, isolated, urban populations of the weed *Crepis sancta* experienced less pollinator activity resulting in less pollination, reduced seed set, and higher selfing rates (Cheptou and Avendaño 2006). Habitat fragmentation can also be a strong selective force against dispersal, leading to the evolution of seed dispersal mechanisms to ensure closer dispersal to the parent plant in isolated, fragmented habitats (Ghazoul 2005). Populations of *Crepis sancta* have significantly more non-dispersing seeds in small, isolated fragments than in larger, continuous habitats (Cheptou et al. 2008). Corridors and stepping stone habitats may be critical to maintaining metapopulation dynamics of plant populations in urban environments, but little is known on the role of corridors in seed dispersal and pollen movement in cities (Van Rossum and Triest 2012).

Stressors of the urban environment on plant communities include modified microclimate; hydrology; soil processes; and air, water, and soil pollution. Soils, in particular, are a driving factor in the establishment and persistence of plant species. Urban soils are highly variable and range dramatically in soil compaction, nutrients, and pollutant content within and across cities (Pavao-Zuckerman 2008; Ossola and Livesley 2016). However, some generalities exist: urban soils are often considered heavily disturbed, with high bulk density, high levels of heavy metals and salinity, and low soil water holding capacity (Pavao-Zuckerman 2008). Properties of urban soils, particularly high nutrient availability, may increase the establishment of invasive plant species (Ehrenfeld 2008). The design or reclamation of urban soils is critical to sustain plant biodiversity (Pavao-Zuckerman 2008) and there are comprehensive guides on designing and managing urban park soils (e.g., Craul and Craul 2006).

Cities are epicenters for non-native species introductions (Pyšek 1998), and vegetation communities in cities are often characterized by novel associations of native and non-native plants (Aronson et al. 2015). When invasive, non-native plants dominate local communities, they often reduce native species growth and diversity (Vilà et al. 2011). Furthermore, recent research has shown that non-native plantings support smaller and less diverse arthropod communities than native plantings, with consequences for populations of animals that depend on arthropod communities as food resources (Burghardt and Tallamy 2013). Reduction and eradication of invasive plants species is a common management technique in remnant natural areas, but invasive plants continue to be planted in private and public UGS. While it is well known by many urban land managers that invasive species are not suitable as planting material in most situations, in some circumstances non-invasive, non-native plant species might be most suitable. For example, in

Toronto, Canada, where there is a bylaw and construction standard governing the implementation of green roofs on most forms of new building, a green roof project can be constructed on new buildings with only 6 cm of growing media if engineers agree the project is suitable given infrastructural constraints. These shallow conditions are not suitable for most plant communities and so non-native species—namely, *Sedum* spp., are primarily used—because of their shallow roots and minimal watering requirements. However, it is critical that managers assess the potential for dispersal, spread, and other mechanisms of invasion before planting non-native species in any type of UGS.

Management decisions, both the introduction, as mentioned above, and removal of particular plant species, by land owners and land managers, are also important determinants of plant community composition and structure in cities. Management activities in UGS are driven not only by federal, municipal, and neighborhood governance, but also by cultural traditions, social norms, socio-economics, and horticultural availability (Ignatieva 2011; Pincetl *et al.* 2013; Grove *et al.* 2014; Aronson *et al.* 2016). This leads to considerable heterogeneity in management practices. Private yards and gardens, in particular, represent the majority of UGS to support urban plant diversity, but each small patch is managed individually. City-wide initiatives that coordinate management activities across public and private UGS could increase connectivity (Goddard *et al.* 2010) and promote metapopulation dynamics necessary for plant population persistence.

Managing urban green spaces for biodiversity conservation

So how do we manage for plant biodiversity given the large variation in biotic and abiotic stressors in different types of urban vegetation communities? Management of vegetation in cities ranges from planting, mowing, and pruning to invasive plant management in remnant natural areas (DiCicco 2014), to large scale restoration projects that require years of planning, implementation, and adaptive management (New York City Department of Parks and Recreation 2014a). The process of identifying, designing, and implementing management activities depends not only on goals related to plant and animal biodiversity, but also on the human use and needs of local communities and other stakeholders. Taking from the ecological restoration literature, management steps can be divided into three stages: 1) planning and design; 2) implementation; and 3) monitoring and adaptive management (New York City Department of Parks and Recreation 2014a; McDonald *et al.* 2016). Planning and design begins with site assessment, including identifying both the needs of UGS users and threats to biodiversity, stakeholder engagement, and setting goals and targets for vegetation management (McDonald *et al.* 2016). Implementation of management activities includes stakeholder engagement and knowledge of when and where local residents use a particular UGS. Implementation of restoration projects in cities entails site protection, invasive plant management, site clearing, soil preparation, and planting activities (New York City Department of Parks and Recreation 2014a).

Finally, monitoring for desired outcomes of management activities is critical, and allows for adaptive management. Here, we present selected case studies for managing urban plant diversity in different types of UGS and discuss the major issues facing urban plant biodiversity and how land managers have dealt with these stressors to maximize successful management outcomes.

Remnant natural areas in cities

Remnant natural areas encompass most of the high quality, native vegetation within urban landscapes. In New York City, for example, 71 percent of all city trees are found within remnant forests (David Nowak, USDA Forest Service, *unpublished data*); thus, providing a disproportionally high amount of benefits such as cooling temperature, absorbing stormwater, and habitat as compared to other types of green spaces within the city. While remnant natural areas contain relatively high native vegetation compared to other types of UGS, vegetation in these natural areas can also contain high proportions of non-native invasive species (Bertin *et al.* 2005). Therefore, restoring degraded remnant areas and sustaining healthy and native vegetation is often a focus of natural resource management within cities.

Suppression and eradication of invasive species can be especially challenging in cities (Rejmanek and Pitcairn 2002; Vidra *et al.* 2007). Invasive species removal projects can be costly without guaranteed outcomes in urban areas, calling to question strategies and targets for vegetation management within highly altered landscapes (Standish *et al.* 2013). To address these challenges, many cities have developed best management practices along with dedicated programs and partnerships (e.g., MillionTreesNYC, Green Seattle Partnership, and Chicago Wilderness). For example, the City of Chicago published the *Chicago Urban Forest Agenda* (2009) with recommended actions for expanding and diversifying native species in natural forests using best management practices. The New York City Department of Parks and Recreation recently published *Guidelines to Urban Forest Restoration* (2014a) which describes methods for forest restoration including invasive removal and native tree planting in natural urban forests in NYC. However, there is still no clear process for defining target conditions and prioritizing where to manage within the urban landscape. With limited budgets, high costs, and high benefits all associated with urban invasive species management, it creates a unique juxtaposition for urban natural resource managers. How can we maintain and manage towards healthy, high functioning forests in a highly stressed and complex urban environment?

In NYC, there are over 117 km^2 of parkland; nearly 35 percent are remnant natural areas comprised of woodlands, grasslands, shrublands, and closed canopy forests (New York City Department of Parks and Recreation 2015). To inform management and establish quantitative vegetation data, a city-wide ecological assessment was conducted in remnant natural areas across the entire city (Forgione *et al.* 2016). Using these data, it was determined that the forest canopy is primarily native (83 percent) but lower proportions of native species were found in the midstory (61 percent) and herbaceous layer (51 percent) (Pregitzer *et al.* in preparation).

Deciding how to prioritize certain species, locations, and/or vegetation layers adds to the complexity of urban natural resource management. For example, Oriental bittersweet (*Celastrus orbiculatus*), a common invasive vine targeted for removal in NYC, was found in 39 percent of all sites evaluated, but varied in percent ground cover from diffuse (0.5 percent) to dominant (83 percent). Focusing on removing this species in sites where its cover is low (< 20 percent) to protect existing healthy native canopy can be achieved with lower intensity work and a lower cost. Once the cover of Oriental bittersweet is high (> 50 percent), it can lead to a homogenous vineland, which can be difficult to maneuver in and requires significantly more resources and time to suppress. With minimal resources, it can be better to work in an area that has mid to low invasive species cover to protect the existing trees and healthy forest than starting an invasive species removal project across a large, highly invaded site.

Appropriately matching economic resources to types of management actions and ecological goals is one strategy to maximize successful outcomes over time. In NYC, removing invasive species and planting native trees and shrubs can be done with different intensities and approaches. Depending on the invasive plant species and extent of its coverage, matching the appropriate type of available workforce is an important step. Examples of the ecological conditions that might be most appropriate for different workforces are:

1. *Hired contractors*, often the most expensive form of management, can be ideal for large project sites and can bring heavy equipment and remove invasive species or debris across large areas. Sites with homogeneous invasive species cover over a large area (i.e. > 2 hectares) that do not contain rare or sensitive species or habitats are ideal for this work.
2. *Technical staff*, crews of natural resource technicians are ideal to work in sensitive ecological areas or sites where heavy equipment is not permitted. They will often have local ecological knowledge and familiarity of parks which can be critical in projects where an invasive species could be easily confused with native species. Because they have continuity with projects they can inform and implement adaptive management.
3. *Volunteer stewards* can be successful in implementing a wide range of projects including invasive species removal, debris or trash removal, trail clearing, and planting native species. It is usually required that sites have safe access, no need for herbicide, motorized equipment, or harmful/toxic plants are present. Staff supervision and training is often required so correct species can be identified and the scope of work is clear. Engaging volunteers through planned events, education and rewards can ensure they continue to engage in future events.

Once invasive species are removed, it must be decided which native species are appropriate to plant. While many factors are considered, in NYC, having broad scale species data from ecological assessments has allowed them to compare the list of species found and improve on their planting palette (Forgione *et al.* 2016).

Next steps of this work are focused on matching what species are present but also expected to thrive in this region under future climate change scenarios (Janowiak *et al.* 2014). More fine resolution data across broad spatial scales is needed for remnant UGS in more cities to inform vegetation patterns, processes, and management. Planting and managing remnant urban areas to be healthy in the face of urban stressors will greatly contribute to overall ecological and social health of cities into the future (Chiesura 2004).

Managed city parks

Cities boast a diverse range of planned and managed park types that vary in their initial conditions, planting strategy, and social programming. Management for plant biodiversity in urban parks may be constrained by both biophysical and social factors. Many designed parks are located in previously developed areas where on-site soil may have been removed and replaced with construction fill (Craul 1992). Existing soil may also be compacted or contaminated from previous land use (Pickett *et al.* 2011). Prior land use and adjacent development may impact site hydrology, increasing stormwater flow and reducing infiltration (Pickett *et al.* 2011). With respect to plant community dynamics, urban parks are often small, edge-dominated patches, embedded within a dense urban matrix and as a result functionally disconnected from other UGS and subject to frequent invasion. Initial and continual plantings are often relied upon as a source of native species. From a social perspective, all urban parks must accommodate human use and when managing for biodiversity designers and managers may be challenged to negotiate between public recreation and ecological conservation (Kowarik and Langer 2005).

Lurie Garden and Brooklyn Bridge Park illustrate the potential for supporting plant biodiversity in the context of constructed and intensely designed, managed, and heavily used urban parks. Lurie Garden, a relatively small 2.0 ha park located within Chicago's Millennium Park, boasts some 222 species of plants, including 20 species of grasses, 26 woody species, and 142 herbaceous perennials. These plantings are separated into two general community types, described as the *dark plate* and *light plate*, representing the tree shaded marshlands and sunny dry prairie of the region, respectively (www.luriegarden.org). Brooklyn Bridge Park (34 ha) is sited on a former industrial park in NYC. Confined to a narrow strip of land located between dense urban settlement and the East River, Brooklyn Bridge Park was designed to accommodate frequent and diverse public use, while integrating multiple habitat types that promote biodiversity and regulate stormwater runoff. The habitat types of the park include ornamental plantings, meadows, shrublands, coastal forest, salt marshes, freshwater wetlands, and lawns (www.brooklynbridgepark.org).

In both Lurie Gardens and Brooklyn Bridge Park, the pedology and hydrology of the sites are landscaped and engineered. Engineered soils are designed, in part, to accommodate stormwater management goals and improve root development and transpiration, and therefore plant health (Bartens *et al.* 2009).

The establishment of soil communities in engineered soils may take time and it is recommended intact soils be preserved whenever possible (New York City Department of Parks and Recreation 2014a; Vergnes *et al.* 2017). However, there is evidence that soil formation and community recovery may be accelerated through the use of soil management interventions, such as amendments and plantings (Vauramo and Setälä 2010; Oldfield *et al.* 2014).

The planting palette of Lurie Garden is not entirely native and instead defined as *near native*, with approximately 40 percent species native to the United States and 25 percent native to the State of Illinois. The goal of this strategy, which blends horticultural and ecological approaches, is to maximize phenologic diversity for aesthetic purposes while supporting urban plant–pollinator interactions by providing abundant and continuous floral resources (Salisbury *et al.* 2015). At Brooklyn Bridge Park, except for the lawns, each microhabitat was planted with native plant species that were sourced from local nurseries as mandated by municipal law (New York City Department of Parks and Recreation 2014b), thereby supporting not just taxonomic, but regional genetic diversity (Zalesny *et al.* 2014).

Lurie Garden employs an adaptive co-management approach to landscape management, which combines multiple stakeholders, different park uses (e.g., social, ecological, aesthetic), and data-driven decision making (Armitage *et al.* 2009). Although the originally designed communities are maintained and succession arrested, natural shifts in species abundance are permitted, unlike other formal gardens. A rotating schedule of mowing regimes has been established and sporadic burnings using hand torches are used to control for invasive grass species. Instead of chemical approaches, plant pests are managed with bio-control strategies. To maximize the success of these strategies and to properly time deployment, plant condition and the populations of insect pests are closely monitored (Crone *et al.* 2009). At Brooklyn Bridge Park, each microhabitat has a unique management prescription that ranges in its intensity and frequency. Recreation spaces are regularly mowed and horticultural zones maintained through regular weeding. In place of burning regimes, meadow patches are mowed once annually to inhibit the establishment of woody species and shrub encroachment. In the coastal woodland patches, natural succession has been allowed to occur, which is evident in the emergence of naturally recruiting tree seedlings, both native and non-native. Given the depauperate available species pool, as is the case with many urban parks, direct planting of shade tolerant species occurs annually to facilitate natural regeneration.

Urban wastelands turned park

Urban parks may also be established on neglected urban spaces, including brownfields and abandoned open space. These areas often contain natural components, including existing vegetation, and as a result may leverage passive restoration strategies as opposed to intensive plantings to promote and maintain biodiversity. Leading examples include Freshkills Park in New York City and Natur-Park Südgelände in Berlin.

Freshkills Park consists of a series of capped landfill mounds, vegetated lowlands, and wetlands, located on the site of the recently decommissioned landfill. The park, which is being opened in phases, extends more than 890 ha—nearly three times the size of Central Park. At Freshkills Park, restoration ecologists have experimented with novel planting strategies to promote native plant establishment and recruitment. One approach is to plant clusters of native trees or shrubs that create perch sites for birds and therefore facilitate natural dispersal of fleshy-fruited plants. Researchers observed that these *nucleated* plantings promote native plant dispersal, depending on the availability of established native plant sources in the area (Robinson and Handel 1993). Recent studies have also sought to select and propagate genotypes of early successional species, including *Salix* spp. and *Populus* spp., that are adapted to the anthropogenic soils found on site (Zalesny *et al.* 2014). Researchers have integrated these *urban genotypes* into restoration plantings as nurse trees, with the intent of accelerating canopy closure and thereby inhibiting the establishment of shade-intolerant invasive species and facilitating the establishment of native woody species recruited through bird dispersal (Richard Hallett, USDA Forest Service, *personal communication*).

Natur-Park Südgelände, an 18 ha former freight rail yard is a leading example of how abandoned urban spaces can be converted to public parks and managed to promote both biodiversity and public use. While we typically associate abandoned urban wastelands with non-native species, such sites have been observed to support diverse flora and fauna, including rare species (Bonthoux *et al.* 2014). Natur-Park Südgelände supports more than 350 plant species, both native and non-native (Kowarik and Langer 2005). As outlined in its management plan, the park employs a simple, direct approach to accommodating public use needs and biodiversity management by defining three basic zones of management: public space, forest, and grassland. Forested areas are left unmanaged as *urban wilderness*, while the grassland, which includes many of the rare species on site, is managed by mowing and grazing to control natural succession. Perhaps what is most unique about the park is that 3.5 ha of the grassland and portions of the forest have been designated as nature conservation areas, where people are confined to the old railways and boardwalks. The design of the conservation areas utilizes the grasslands as view-sheds into the untamed wild, engaging with public perceptions of fear and curiosity associated with urban natural areas (e.g., Jim and Chen 2006; Lyytimäki *et al.* 2008).

Green infrastructure

Green infrastructures, including green roofs and bioswales, were traditionally designed to capture stormwater to reduce the volume and flow rate reaching sewer treatment facilities in cities and towns. Recently, however, there is an increasingly compelling link between the vegetation community type and diversity of plant communities on these green infrastructures, and the different ecosystem services provided by them (Lundholm 2015; MacIvor *et al.* 2016). These include the management of rainwater (Lundholm *et al.* 2010), building cooling (Blanusa *et al.* 2013), habitat for urban biodiversity (Madre *et al.* 2014; Williams *et al.* 2014), and aesthetic value (Lee *et al.* 2014).

Green infrastructures are exposed to environmental conditions that present extreme habitat for plants to survive. For example, green roofs are on top of buildings where they may experience full sun and added reflection from nearby windows, high winds and wind scour off buildings, as well as other microclimate factors not typical of ground level UGS. Supplemental watering is often needed to manage green roof vegetation that experiences drought due to exposure and shallow substrate. Green roofs are generally far windier than habitats at ground level protected from buildings. During high winds, green roofs can experience uplift and damage to waterproofing membranes and substrates become susceptible to blowing off of the roof, exposing plant roots in the process—leading to higher rates of mortality (Dunnett and Kingsbury, 2008). A plastic-based stabilization mesh is sometimes laid over the plants temporarily to reduce this damage until the roots have sufficiently bound into the growing media.

These conditions have resulted in a reduced list of plant species that are suitable for the most commonly specified green roofs worldwide, which consist of shallow low organic substrates of less than 15 cm (termed *extensive green roofs*; Snodgrass and Snodgrass 2006). These extensive green roof plant communities are dominated by a group of succulent CAM plants from the genus *Sedum* that are widely available in the horticultural industry and outlive other plant communities in comparative studies. For example, in Michigan, Monterusso *et al.* (2005) examined a number of native meadowland perennials and *Sedum* on extensive green roofs and found that after exposure to drought conditions only the *Sedum* species survived. In other regions that are cooler and wetter, such as Nova Scotia, plant communities that are comprised of different plant species, including *Sedum*, but also grasses and perennial flowering plants, outperformed any other combination of plant species consisting of two or less of these three plant groups (Lundholm *et al.* 2010). Although in many regions outside of Europe, and especially in North America, the majority of *Sedum* are non-indigenous species, they continue to be the predominantly used plant group because they are easily cultivated, installed, and can withstand significant exposure (Rowe *et al.* 2012).

Similarly, bioswales and other green infrastructure plantings adjacent to hard surfaces designed to infiltrate stormwater experience extreme conditions including inundation of water following rain events punctuated by periods of extreme drought. These water retaining green infrastructures are also exposed to higher concentrations of salt, pollutants, and other toxic materials that accumulate on roads and are washed into bioswales during rain events and after snow melting (Barrett *et al.* 1998). The plant community that is specified for these retention areas must be tolerant of halophytic conditions and periods of standing water and so there is a limited set of species that are conventionally specified.

Cultivating a specific plant community composition rather than simply maximizing plant cover in order to increase ecosystem services requires specialized knowledge relating to plant identification at different life stages and selective weeding. Of course no two green infrastructure projects are alike, and so it is critical to interpret which microclimate characteristics will present management issues that are site specific and can be planned for accordingly. Incorporating these characteristics into the initial design will minimize uncertainty and costs associated with cultivating a more native and diverse plant community. More work is needed to test and research different plant species, traits, and community combinations suitable for green roof applications (Lundholm and Williams 2015). More broadly, this approach is needed for all green infrastructures (MacIvor *et al.* 2016), and this research should integrate multiple actors including the horticultural industry, ecologists, landscape architecture, and other practitioners that specialize in management of green infrastructure.

Specialized knowledge is necessary for practitioners tasked with managing the plant communities living on green infrastructures. This includes specialized gardening tools and the safety of the maintenance staff. Green infrastructures occur in specialized locations and are often integrated into other purpose built structures such as building roofs, walls, or along roadsides where other utilities are shared, such as telecommunications, sewage, and water mains. Tie-offs, harnesses, and fall arrest training are often requirements of green roof management practitioners (Behm 2011). In another example, along roadsides where bioswales and retention ponds are normally integrated (Xiao and McPherson 2011), maintenance staff must be aware of moving vehicles and potential contact with contaminants. Management of plant communities on green infrastructures requires careful planning, training, and equipment for successful outcomes.

Managing for humans and nature

A primary challenge for management of plant communities in UGS are tradeoffs between human use and management for biodiversity. Human activity can result in trampled vegetation or compacted soil, affecting vegetation and inhibiting natural recruitment. For UGS that include natural areas, successful strategies for preserving plant biodiversity include clearly defining spaces for active recreation and buffering natural areas with plantings and trail networks that allow people to move through, but not access certain areas of natural habitat. Urban green space managers must also accommodate a range of social perceptions to natural areas. Research has observed dense vegetation to be associated with fear and concerns of safety (Lyytimäki *et al.* 2008). In such instances the appropriate response may be to open sightlines and reduce planting structure (Kuo and Sullivan 2001)—a strategy that may reduce plant biodiversity. However, other studies have found complex vegetation structure and composition are preferred (Jim and Chen 2006) and promote the sense of *wilderness* (e.g. Natur-Park Südgelände; Kowarik and Langer 2005). Another challenge is communicating the intent of management strategies in natural areas, which may be interpreted as messy, disorderly, or unkempt. Ultimately, there is a need to recognize these social interactions and integrate ecological management in UGS with novel design strategies, signage, civic engagement and community participation (Nassauer 2013). The role of community engagement and stewardship in the promotion of plant diversity, structure, and health, may be even more important in less formal UGS, including streetscapes and pocket parks (Jack-Scott *et al.* 2013).

Ultimately, the true success of managing UGS for plant biodiversity will be in recognizing these interactions between people and urban plants and strengthening support for research, design, management, and civic engagement. For this to occur, however, collaborations among urban natural resource managers, design practitioners, and urban planners with ecological and social scientists are essential. Mechanisms that provide opportunities for these collaborations, such as dedicated funding from city and federal governments and local and global networks that promote such interdisciplinary research are needed. Clear targets for monitoring, research, and adaptive management activities for urban plant biodiversity are only possible with these collaborations.

References

Ahern J., Cilliers S.S. and Niemelä J. (2014) The concept of ecosystem services in adaptive urban planning and design: A framework for supporting innovation. *Landscape and Urban Planning* 125: 254–259.

Armitage D.R., Plummer R., Berkes F., Arthur R.I., Charles A.T., Davidson-Hunt I.J., Diduck A.P., Doubleday N.C., Johnson D.S., Marschke M., McConney P., Pinkerton E.W. and Wollenberg E.K. (2009) Adaptive co-management for social-ecological complexity. *Frontiers in Ecology and the Environment* 7: 95–102.

Aronson M.F.J., La Sorte F.A., Nilon C.H., Katti M., Goddard M.A., Lepczyk C.A., Warren P.S., Williams N.S.G., Cilliers S.S., Clarkson B., Dobbs C., Dolan R., Hedblom M., Klotz S., Louwe Kooijmans J., Kühn I., MacGregor-Fors I., McDonnell M., Mörtberg U., Pyšek P., Siebert S., Sushinsky J., Werner P. and Marten W. (2014) A global analysis of the impacts of urbanization on bird and plant diversity reveals key anthropogenic drivers. *Proceedings of the Royal Society B Biological Sciences* 281: 20133330.

Aronson M.F.J., Handel S.N., La Puma I.P. and Clemants S.E. (2015) Replacement of native communities with novel plant assemblages dominated by non-native species in the New York metropolitan region. *Urban Ecosystems* 18: 31–45.

Aronson M.F.J., Nilon C.H., Lepczyk C.A., Parker T.S., Warren P.S., Cillier S.S., Goddard M.A., Hahs A.K., Herzog C., Katti M., La Sorte F.A., Williams N.S.G. and Zipperer W. (2016) Hierarchical filters determine community assembly of urban species pools. *Ecology* 97: 2952–2963.

Aronson M.F.J., Lepczyk C.A., Evans K.L., Goddard M.A., Lerman S.B., MacIvor J.S., Nilon C.H. and Vargo T. (2017) Biodiversity in the city: Key challenges for urban green space management. *Frontiers in Ecology and the Environment* 15: 189–196.

Ballard M., Hough-Goldstein J. and Tallamy D. (2013) Arthropod communities on native and nonnative early successional plants. *Environmental Entomology* 42: 851–859.

Barrett M.E., Walsh P.M., Malina Jr J.F. and Charbeneau R.J. (1998) Performance of vegetative controls for treating highway runoff. *Journal of Environmental Engineering* 124: 1121–1128.

Bartens J., Day S.D., Harris J.R., Wynn T.M. and Dove J.E. (2009) Transpiration and root development of urban trees in structural soil stormwater reservoirs. *Environmental Management* 44: 646–657.

Behm M. (2011) Safe design suggestions for vegetated roofs. *Journal of Construction Engineering and Management* 138: 999–1003.

Beninde J., Veith M. and Hochkirch A. (2015) Biodiversity in cities needs space: a meta-analysis of factors determining intra-urban biodiversity variation. *Ecology Letters* 18: 581–92.

Bertin R.I., Manner M.E., Larrow B.F., Cantwell T.W. and Berstene. E.M. (2005) Norway maple (*Acer platanoides*) and other non-native trees in urban woodlands of central Massachusetts. *The Journal of the Torrey Botanical Society* 132: 225–235.

Blanusa T., Monteiro M.M.V., Fantozzi F., Vysini E., Li Y. and Cameron R.W. (2013) Alternatives to Sedum on green roofs: Can broad leaf perennial plants offer better 'cooling service'? *Building and Environment* 59: 99–106.

Bonthoux S., Brun M., Di Pietro F., Greulich S. and Bouché-Pillon S. (2014) How can wastelands promote biodiversity in cities? A review. *Landscape and Urban Planning* 132: 79–88.

Burghardt K. and Tallamy D.W. (2013) Plant origin asymmetrically impacts feeding guilds and life stages driving community structure of herbivorous arthropods. *Diversity and Distributions* 19: 1553–1565.

Cheptou P.-O. and Avendaño V.L.G. (2006) Pollination processes and the Allee effect in highly fragmented populations: Consequences for the mating system in urban environments. *New Phytologist* 172: 774–783.

Cheptou P.-O., Carrue O., Rouifed S. and Cantarel A. (2008) Rapid evolution of seed dispersal in an urban environment in the weed *Crepis sancta*. *Proceedings of the National Academy of Sciences* 105: 3796–3799.

Chiesura A. (2004) The role of urban parks for the sustainable city. *Landscape and Urban Planning* 68: 129–138.

Cilliers S., Cilliers J., Lubbe R. and Siebert S. (2013) Ecosystem services of urban green spaces in African countries: Perspectives and challenges. *Urban Ecosystems* 16: 681–702.

City of Chicago (2009). Chicago's Urban Forest Agenda. Available at www.cityofchicago.org/content/dam/city/depts/doe/general/NaturalResourcesAndWaterConservation_PDFs/UrbanForestAgenda/ChicagosUrbanForestAgenda2009.pdf (accessed 5 February 2017).

Craul P.J. (1992) *Urban Soil in Landscape Design*. John Wiley & Sons, New York.

Craul T.A. and Craul P.J. (2006) *Soil Design Protocols for Landscape Architects and Contractors*, John Wiley & Sons, New York.

Crone E.E., Marler M. and Pearson D.E. (2009) Non target effects of broadleaf herbicide on a native perennial forb: A demographic framework for assessing and minimizing impacts. *Journal of Applied Ecology* 46: 673–682.

DiCicco J.M. (2014). Long-term urban park ecological restoration: A case study of Prospect Park, Brooklyn, New York. *Ecological Restoration* 32: 314–326.

Dunnett N. and Kingsbury N. (2008) *Planting Green Roofs and Living Walls*. Timber Press, Portland, OR.

Ehrenfeld J.G. (2000) Evaluating wetlands within an urban context. *Ecological Engineering* 15: 253–265.

Forgione H.M., Pregitzer C.C., Charlop-Powers S. and Gunther B. (2016) Advancing urban ecosystem governance in New York City: Shifting towards a unified perspective for conservation management. *Environmental Science & Policy* 62: 127–132.

Fuller R.A., Irvine K.N., Devine-Wright P., Warren P.H. and Gaston K.J. (2007) Psychological benefits of greenspace increase with biodiversity. *Biology Letters* 3: 390–394.

Ghazoul J. (2005) Pollen and seed dispersal among dispersed plants. *Biological Reviews* 80(3): 413–443.

Gibbs J.P. (2000) Wetland loss and biodiversity conservation. *Conservation Biology* 14: 314–317.

Goddard M.A., Dougill A.J. and Benton T.G. (2010) Scaling up from gardens: Biodiversity conservation in urban environments. *Trends in Ecology and Evolution* 25: 90–98.

Gómez-Baggethun E., Gren Å., Barton D.N., Langemeyer J., McPhearson T., O'Farrell P., Andersson E., Hamstead Z. and Kremer P. (2013) Urban ecosystem services, in Elmqvist

T., Fragkias M., Goodness J., Güneralp B. and Marcotullio, P.J. (eds), *Urbanization, Biodiversity and Ecosystem Services: Challenges and Opportunities: A Global Assessment*, pp. 175–251. Springer, Netherlands.

Grove J.M., Locke D.H. and O'Neil-Dunne J.P. (2014) An ecology of prestige in New York City: Examining the relationships among population density, socio-economic status, group identity, and residential canopy cover. *Environmental Management* 54: 402–419.

He C., Liu Z., Tian J. and Ma Q. (2014) Urban expansion dynamics and natural habitat loss in China: A multiscale landscape perspective. *Global Change Biology* 20: 2886–2902.

Hennig E.I. and Ghazoul J. (2011) Plant–pollinator interactions within the urban environment. *Perspectives in Plant Ecology, Evolution and Systematics* 13: 137–150.

Ignatieva M. (2011) Plant material for urban landscapes in the era of globalisation: Roots, challenges, and innovative solutions, in M. Richter and U. Weiland (eds), *Applied Urban Ecology: A Global Framework*, pp. 139–151. Blackwell, Hoboken, NJ.

Ives C.D., Lentini P.E., Threlfall C.G., Ikin K., Shanahan D.F., Garrard G.E., Bekessy S.A., Fuller R.A., Mumaw L., Rayner L., Rowe R., Valentine L.E. and Kendal D. (2016) Cities are hotspots for threatened species. *Global Ecology and Biogeography* 25: 117–126.

Jack-Scott E., Piana M., Troxel B., Murphy-Dunning C. and Ashton M.S. (2013) Stewardship success: How community group dynamics affect urban street tree survival and growth. *Arboriculture and Urban Forestry* 39: 189–196.

Janowiak M.K., Swanston C.W., Nagel L.M., Brandt L.A., Butler P.R., Handler S.D., Shannon P.D., Iverson L.R., Matthews S.N., Prasad A. and Peters M.P. (2014) A practical approach for translating climate change adaptation principles into forest management actions. *Journal of Forestry* 112: 424–433.

Jim C.Y. and Chen W.Y. (2006) Perception and attitude of residents toward urban green spaces in Guangzhou (China). *Environmental Management* 38: 338–349.

Kowarik I. and Langer A. (2005) Natur-Park Südgelände: Linking conservation and recreation in an abandoned railyard in Berlin. In Kowarik I. and Körner S. (eds), *Wild Urban Woodlands*, pp. 287–299. Springer, Berlin, Heidelberg.

Kuo F.E. and Sullivan W.C. (2001) Environment and crime in the inner city: Does vegetation reduce crime? *Environment and Behavior* 33: 343–367.

Lee K.E., Williams K.J., Sargent L.D., Farrell C. and Williams N.S. (2014) Living roof preference is influenced by plant characteristics and diversity. *Landscape and Urban Planning* 122: 152–159.

Lubbe C.S., Siebert S.J. and Cilliers S.S. (2010) Political legacy of South Africa affects the plant diversity patterns of urban domestic gardens along a socio-economic gradient. *Science Research Essays* 5: 2900–2910.

Lundholm J., MacIvor J.S., MacDougall Z. and Ranalli M. (2010) Plant species and functional group combinations affect green roof ecosystem functions. *Plos One* 5: e9677.

Lundholm J. and Williams N.S. (2015) Effects of vegetation on green roof ecosystem services. In Sutton R.K. (ed.), *Green Roof Ecosystems*, pp. 211–232. Springer International Publishing.

Lundholm J.T. (2015) Green roof plant species diversity improves ecosystem multifunctionality. *Journal of Applied Ecology* 52: 726–734.

Lyytimäki J., Petersen L.K., Normander B. and Bezák P. (2008) Nature as a nuisance? Ecosystem services and disservices to urban lifestyle. *Environmental Sciences* 5: 161–172.

MacIvor J.S., Cadotte M.W., Livingstone S.W., Lundholm J.T. and Yasui S.L.E. (2016) Phylogenetic ecology and the greening of cities. *Journal of Applied Ecology* 53: 1470–1476.

Madre F., Vergnes A., Machon N. and Clergeau P. (2014) Green roofs as habitats for wild plant species in urban landscapes: First insights from a large-scale sampling. *Landscape and Urban Planning* 122: 100–107.

McDonald T., Gann G.D., Jonson J. and Dixon K.W. (2016) *International Standards for the Practice of Ecological Restoration – Including Principles and Key Concepts.* Society for Ecological Restoration, Washington, DC.

Monterusso M.A., Rowe D.B. and Rugh C.L. (2005) Establishment and persistence of Sedum spp. and native taxa for green roof applications. *HortScience* 40: 391–396.

Nassauer J. (Ed.) (2013) *Placing Nature: Culture and Landscape Ecology,* Island Press, Washington, DC.

New York City Department of Parks and Recreation (2014a) Guidelines for Urban Forest Restoration. NYC Parks, Natural Resources Group, New York. Available at www. nycgovparks.org/pagefiles/84/guidelines-to-urban-forest-restoration.pdf (accessed 5 February 2017).

New York City Department of Parks and Recreation (2014b) Native Species Planting Guide for New York City. NYC Parks, Natural Resource Group, New York. Available at www.nycgovparks.org/pagefiles/92/nrg-native-species-planting-guide.pdf (accessed 5 February 2017).

New York City Department of Parks and Recreation (2015) About NYC Parks. NYC Parks, New York. Available at www.nycgovparks.org/about (accessed 5 Feburary 2015).

Oldfield E.E., Felson A.J., Wood S.A., Hallett R.A., Strickland M.S. and Bradford M.A. (2014) Positive effects of afforestation efforts on the health of urban soils. *Forest Ecology and Management* 313: 266–273.

Ossola, A., Livesley, S.J., 2016. Drivers of soil heterogeneity in the urban landscape. In Francis, R.A., Millington, J., Chadwick, M.A. (eds), Urban *Landscape Ecology: Science, Policy and Practice,* pp. 19–41. Routledge, New York.

Pavao-Zuckerman M.A. (2008) The nature of urban soils and their role in ecological restoration in cities. *Restoration Ecology* 16: 642–649.

Pickett S.T.A., Cadenasso M.L., Grove J.M., Boone C.G., Groffman P.M., Irwin E., Kaushal S.S., McGrath B.P., Nilon C.H., Pouyat R.V., Szlavecz K., Troy A. and Warren P. (2011) Urban ecological systems: Scientific foundations and a decade of progress. *Journal of Environmental Management* 92: 331–362.

Pincetl S., Prabhu S.S., Gillespie T.W., Jenerette G.D. and Pataki D.E. (2013) The evolution of tree nursery offerings in Los Angeles County over the last 110 years. *Landscape and Urban Planning* 118: 10–17.

Pregitzer C.C., Bibbo S., Forgione H.M., Hallett R.A., Charlop-Powers S. and Bradford M.A. (in preparation). Ecological Complexity of Human Altered Landscapes. Fine scale data drives patterns of native species composition in New York City forests.

Pyšek P. (1998) Alien and native species in central European floras: A quantitative comparison. *Journal of Biogeography* 25:155–163.

Rebelo A.G., Holmes P.M., Dorse C. and Wood J. (2011) Impacts of urbanization in a biodiversity hotspot: Conservation challenges in Metropolitan Cape Town. *South Africa Journal of Botany* 77: 20–35.

Rejmanek M. and Pitcairn M.J. (2002) When is eradication of exotic pest plants a realistic goal? In Veitch C.R. and Clout M.N. (eds), *Turning the Tide: The Eradication of Invasive Species,* pp. 249–253. Auckland, New Zealand: Invasive Species Specialist Group of the World Conservation Union (IUCN).

Robinson G.R. and Handel S.N. (1993) Forest restoration on a closed landfill: Rapid addition of new species by bird dispersal. *Conservation Biology* 7: 271–278.

Rowe D.B., Getter K.L. and Durhman A.K. (2012) Effect of green roof media depth on Crassulacean plant succession over seven years. *Landscape and Urban Planning* 104: 310–319.

Salisbury A., Armitage J., Bostock H., Perry J., Tatchell M. and Thompson K. (2015) Enhancing gardens as habitats for flower-visiting aerial insects (pollinators): Should we plant native or exotic species? *Journal of Applied Ecology* 52: 1156–1164.

Snodgrass E.C. and Snodgrass L.L. (2006) *Green Roof Plants: A Resource and Planting Guide.* Timber Press, Portland, OR.

Standish R.J. Hobbs R.J. and Miller J.R. (2013) Improving city life: Options for ecological restoration in urban landscapes and how these might influence interactions between people and nature. *Landscape Ecology* 28: 1213–1221.

Threlfall C.G., Mata L., Mackie J.A., Hahs A.K., Stork N.E., Williams N.S.G. and Livesley S.J. (2017) Increasing biodiversity in urban green spaces through simple vegetation interventions. *Journal of Applied Ecology* Online First DOI: 10.1111/1365-2664.12876.

Van Rossum F. and Triest L. (2012) Stepping-stone populations in linear landscape elements increase pollen dispersal between urban forest fragments. *Plant Ecology and Evolution* 145: 332–340.

Vauramo S. and Setälä H. (2010) Urban belowground food-web responses to plant community manipulation: Impacts on nutrient dynamics. *Landscape and Urban Planning* 97: 1–10.

Vergnes A., Blouin M., Muratet A., Lerch T.Z., Mendez-Millan M., Rouelle-Castrec M. and Dubs F. (2017) Initial conditions during Technosol implementation shape earthworms and ants diversity. *Landscape and Urban Planning* 159: 32–41.

Vidra R.L., Shear T.H., Stucky J.M. (2007) Effects of vegetation removal on native understory recovery in an exotic-rich urban forest. *Journal of the Torrey Botanical Society* 134: 410–419.

Vilà M., Espinar J.L., Hejda M., Hulme P.E., Jarošik V., Maron J.L., Pergl J., Schaffner U., Sun Y. and Pyšek P. (2011) Ecological impacts of invasive alien plants: A meta-analysis of their effects on species, communities and ecosystems. *Ecology Letters* 14: 702–708.

Williams N.S., Lundholm J. and MacIvor J.S. (2014) Do green roofs help urban biodiversity conservation? *Journal of Applied Ecology* 51: 1643–1649.

Williams N.S.G., Schwartz M.W., Vesk P.A., McCarthy M.A., Hahs A.K., Clemants S.E., Corlett R.T., Duncan R.P., Norton B.A., Thompson K. and McDonnell M.J. (2009) A conceptual framework for predicting the effects of urban environments on floras. *Journal of Ecology* 97: 4–9.

Xiao Q. and McPherson E.G. (2011) Performance of engineered soil and trees in a parking lot bioswale. *Urban Water Journal* 8: 241–253.

Zalesny R.S., Hallett R.A., Falxa-Raymond N., Wiese A.H. and Birr B.A. (2014) Propagating native Salicaceae for afforestation and restoration in New York City's five boroughs. *Native Plants Journal* 15: 29–41.

8

MANAGING BIODIVERSITY THROUGH SOCIAL VALUES AND PREFERENCES

Kate E. Lee and Dave Kendal

Introduction

Managing biodiversity in cities effectively requires the combined understanding of a number of domains across the social and natural sciences. Social researchers, in particular, play a crucial part in shaping policy, planning and management decisions for urban greenspaces from a local to international scale (Williams and Cary 2002). The relentless march of urbanization through sprawl and densification, and associated loss of biodiversity worldwide, raises questions about the importance of managing urban greenspaces in ways that promote biodiversity and encourage biodiversity conservation (Nielsen *et al.* 2013). Biodiverse greenspaces provide many benefits for people by providing opportunities to interact with nature (Dallimer *et al.* 2012; Shwartz *et al.* 2014) and for recreation. Less clear, however, is how biodiversity influences people's perceptions and experiences of nature, and in turn, how much they value this experience (Voigt and Würster 2015).

This chapter will provide an overview of research exploring people's experiences related to biodiversity. It will set out a framework for understanding how people perceive and respond to biodiversity in the landscape, first outlining the role of environment-related factors such as landscape features and context, and people-related factors such as values and preference, before briefly touching on implications for biodiversity conservation behaviours. The chapter will finish by providing suggestions for managing different kinds of urban landscapes in ways that support both people and the provision of biodiversity.

Can people accurately perceive biodiversity?

Understanding how people will respond to and appreciate biodiversity first requires an understanding of how people perceive, or are aware of, biodiversity. There is currently a lack of consensus on this topic, underscored by contradictory findings between studies (e.g. Fuller *et al.* 2007; Dallimer *et al.* 2012) and within studies (e.g. Fuller *et al.* 2007). For example, research by Fuller and colleagues (2007) showed that people were able to accurately perceive plant biodiversity, but not bird or butterfly biodiversity. Further research by Shwartz and colleagues (2014) showed that in fact, people may underestimate flower, bird and pollinator biodiversity by as much as 50 per cent. People's reports of biodiversity are most inaccurate at extremes of the biodiversity spectrum; they consistently overestimate biodiversity when it is low and underestimating biodiversity when it is high. However, there is some evidence that accuracy is improved with higher levels of planting evenness, and with environmental expertise.

There are likely to be a range of reasons that experts differ from non-experts in their perceptions of biodiversity. For example, non-experts often look for proxy markers of biodiversity such as obvious or charismatic animals, vegetation diversity or vegetation structural heterogeneity (Qiu *et al.* 2013; Shwartz *et al.* 2014). Members of the public may think of biodiversity differently than experts and scientists (Shwartz *et al.* 2014; Voigt and Würster 2015). Research suggests that they may be less concerned with naming particular species or features of environments and more concerned with the way that these species and features seem to fit with each other and with the local context (Voigt and Würster 2015). To better understand how people think about biodiversity, Voigt and Würster (2015) conducted qualitative interviews with 76 day visitors to a study site in Salzburg, Austria. Interviewees correctly perceived high levels of biodiversity at the study site and reported being confident in their biodiversity assessment. But when prompted for more details of the biodiversity, interviewees provided few details about species or landscape structures. Instead, reports of biodiversity reflected people's responses to their *experience* in the environment and the benefits they perceived to gain from that. This suggests that for certain people at least, perceptions of biodiversity may take place in a holistic manner, taking into account people's perceived experiences in that environment.

Experts and non-experts, however, both draw on their underlying knowledge to extrapolate estimates of biodiversity (Qiu *et al.* 2013) and so increasing biodiversity awareness in non-experts could help them more accurately perceive biodiversity (Shwartz *et al.* 2012). Providing information on biodiversity may help to boost people's awareness of, and appreciation for biodiversity, and in turn, their willingness to conserve it (Shwartz *et al.* 2012; 2014). To test this, Shwartz and colleagues (2014) ran a large-scale field study over one year involving 1116 people within 14 inner-city gardens in Paris, France. Half of these gardens changed to biodiversity-friendly practices which increased their level of biodiversity and half of these gardens in turn provided additional information about this change. Information was provided in the form of signage, through the local gardeners, and in some cases, activity days. But despite providing information about the significant increases in biodiversity, members of the public continued to underestimate species richness, only providing accurate estimates for native flowers. Interestingly, however, the authors report a *people-biodiversity paradox* whereby members of the public reported that they liked rich species diversity and felt it important for their well-being, even though they did not appear to notice the changes in biodiversity induced by the researchers (Shwartz *et al.* 2014).

To further complicate this discussion, research suggests that accurate perceptions of biodiversity are not necessarily correlated with aesthetic preferences. Research by Qiu and colleagues (2013) showed discrepancies between biodiversity percep-tions and ratings of aesthetic preferences; low biodiversity ornamental landscapes were often highly preferred, while high biodiversity forest landscapes were less preferred. In making their assessments of biodiversity, participants focused pre-dominantly on vegetation, possibly because it comprises a large proportion of the landscape and because it is static. The authors identified a number of factors besides biodiversity that they felt may have influenced aesthetic preference ratings: in the wilder, more biodiverse landscapes, people tended to react very negatively to signs of human presence. For example, there were many negative comments around fences and bridges because respondents felt that they didn't fit with the landscape, but rather detracted from it (Qiu *et al.* 2013). Again, perceptions of the fit between biodiversity and local context appear important where they are presumed to reflect respect for the environment (Sheppard 2001) (Figure 8.1).

Do people prefer biodiversity?

A large body of literature has explored people's preferences for landscapes, and how different physical features of these landscapes influence preference for land-scape (e.g. Kendal *et al.* 2012a; Lee *et al.* 2014). Preferences, positive evaluations, and feelings expressed as liking, are often used to better understand how to design and manage urban greenspaces in ways that the public will appreciate (Kaplan *et al.* 1998).

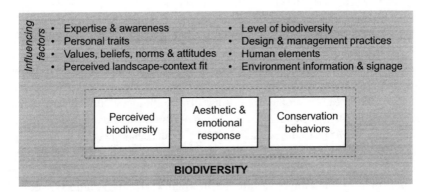

FIGURE 8.1   A conceptual model highlighting the role of people–environment factors in influencing biodiversity, aesthetics and emotional responses, and conservation behaviours.

There are a range of underlying theories that have been used to explain these effects, depending on whether people are presumed to respond to the landscape in ways that are evolutionary adaptations, as a source of information to be processed, or according to normative social–cultural processes. Various evolutionary preference theories state that some landscapes may be preferred because the human species evolved to prefer landscapes that provided good habitat (Orians and Heerwagen 1992). That is, landscapes provided some features that enhanced survival for our ancestors, and adaptations to prefer these features were selected through evolution. These theories highlight preferences for savanna landscapes with *easy-to-traverse* grassy understories with scattered trees with spreading canopies to provide protection from predators and provide shade (Orians and Heerwagen 1992). As well as providing protective refuges, they provided prospect opportunities for scouting the landscape (Appleton 1975) and included features such as lush green foliage, water and brightly coloured flowers or berries (Orians and Heerwagen 1992).

A related theory sees humans as information processors and the external environment as a source of information to be processed, with certain environments being easier to process than others. In this instance, preference reflects landscape features and spatial organization that is quicker and easier to understand, and therefore, to process. This theory states that people prefer moderately *complex* landscapes that are not too boring or too confusing, that are organized into *coherent* groupings, that are *legible* and perceived as easy to navigate, and that are *mysterious* and so encourage a desire for greater exploration (Kaplan and Kaplan 1989). These factors of complexity, coherence, legibility and mystery may interact with each other such that, for example, preferences for highly complex landscapes are enhanced with correspondingly high levels of coherence (Kaplan et al. 1998).

A final cluster of theories assigns a greater role to social and cultural processes in shaping landscape preferences. Like the previous theory, they too consider the external environment as a source of information, but here landscape information is used to communicate with other people. For example, studies by Nassauer show a general preference in developed Western countries for *cues of care* in landscapes, shaped by normative influences for neatness (e.g. Nassauer 1995). In this case, cues of care such as fences and mown edges provide options for presenting messy eco-systems in ways that are preferred (Nassauer 1995). Specific investigations on the effect of cultural groupings on preferences for biodiverse landscapes have shown that there are some small but significant differences. Members of a local wild-flower group in the state of Western Australia had distinctly different preferences for local natural landscapes than both a group of Australian students and a group of American students (Kaplan and Herbert 1987). Farmers have been shown, through a series of studies in the Netherlands, to have lower preferences for natural, compared to managed landscapes (e.g. van den Berg et al. 2006). Interestingly, several studies based on the ethnic background have shown relatively little variation in landscape preference (e.g. Yang and Kaplan 1990).

Biodiversity may fit within these frameworks in different ways. For example, through the lens of evolutionary theories, preferences for biodiversity as a stable, healthy environment that provides a range of materials to enhance survival could be interpreted as a result of its providing fitness advantages for our ancestors. Through the lens of information processing theory, biodiversity is an important driver of complexity and preferences might follow this relationship (complexity implicitly refers to *perceived* biodiversity and is often applied to vegetation characteristics alone). Lastly, through the lens of socio-cultural theories, preferences for biodiversity could reflect the extent to which the presence of biodiversity is socially accepted, or the extent to which it is managed in ways that reflect social norms (e.g. cues for care).

These theories largely rely on visual perception. Recently, however, researchers have considered the role of other senses, such as sound and smell, in shaping nature experiences. Most of this research has been focused on the role of these senses in influencing health and well-being (see Chapter 9 for more detail on health and well-being), so their role in shaping preferences is less clear. Landscape perception is likely to be a multi-sensory experience as information from multiple senses is integrated, along with our existing understandings and connections with the environment, to inform our impression of the external landscape (Bundy *et al.* 2002). This suggests that different people may perceive and experience biodiversity in different ways.

Other factors influencing perceptions of, and preferences for, biodiversity

Cognitive factors

A cognitive approach to understanding people's reactions to natural environments suggests that different people will have different preferences and perceptions of the same landscape based on their values, beliefs, norms and attitudes. Cultural theories of landscape preference are less likely to look for, and see preference as a common consensus and instead, consider differences in the way that different groups of people respond to landscapes. A variety of studies have shown that values can influence landscape preference (Bjerke *et al.* 2006). People with an anthropocentric, or people-centred, view of the relationship between people and nature preferred overtly cultural landscapes, while people with ecocentric, or nature-centred, views tended to prefer wilder landscapes (de Groot and van den Born 2003). Preference for wild landscapes has also been related to personal traits; people with higher levels of personal need for structure preferred wild landscapes less and manicured landscapes more than people with lower levels of this trait (van den Berg and van Winsum-Westra 2010). Differences in preferences between experts (including landscape architects and arborists) and the general public are likely shaped by underlying cognitive differences in their values and beliefs.

Another important implication of cultural and cognitive theories is that they suggest that people's preferences are malleable. This contrasts with evolutionary theories that see landscape preference as an expression of genetic pre-determinants. Gobster (1999) has argued that people tend to judge landscapes using a scenic aesthetic mode, largely based on the aesthetic properties (such as beauty) of a scene, unless they know about deeper ecological functioning of a landscape (e.g. through the provision of information) that allows them switch to using an ecological aesthetic mode, where judgments are also based on ecological properties such as the biodiversity value of a landscape (Gobster, 1999). Recent studies suggest that the provision of information can indeed influence preference for biodiverse landscapes where this information is consistent with people's values (Straka *et al.* 2016).

Demographic factors

Socio-economics is often considered an important driver of urban biodiversity. In Phoenix, USA, research has shown that urban biodiversity is strongly related to socio-economic status – areas where people with high incomes live have higher levels of biodiversity (species richness) than areas where people with lower levels of income live (Hope *et al.* 2003). This has been attributed to a *luxury effect*, where people with higher incomes have the economic wherewithal to move to places with higher levels of biodiversity. However, others have shown that these patterns do not occur in all urban areas (Cilliers and Siebert 2011; Kendal *et al.* 2012b). An alternative theory about the distribution of vegetation is that unequal power relations between different parts of the community and centralized management authorities leads to unequal provision of a public good (parks and street trees) for the private benefits of advantaged members of the community. There is some empirical evidence on tree cover that supports this (Kendal *et al.* 2012b).

Demographic factors such as household income, education level, age and gender are commonly found to be significant but minor predictors of a range of attitudes related to biodiversity, such as preference (van den Berg *et al.* 2006) and acceptability of management practices (Kleiven *et al.* 2004). Sometimes mechanistic explanations are used to explain these findings; for example, higher education levels are used as a proxy for knowledge about biodiversity. However, it is likely that cognitive factors are stronger predictors and more closely related to the mechanisms that underpin variation in preference for biodiversity.

Beyond preferences

Valuing biodiversity

The concept of preference also has a particular interpretation in economics, where it refers to a preferred order among a set of alternatives. Preferences in this sense are an important influence on the choices people make, and particularly important in determining the *value* people place on things. These conceptions of preference and value have been widely adopted in the ecosystem services discourse (e.g. Gomez-Baggethun and Barton 2013) and by biodiversity conservation more generally (e.g. Dearborn and Kark 2009). The value *of* biodiversity is considered commensurable; able to be converted into a single measure of value (typically money).

This approach has led to the development of a number of models to calculate the value of biodiversity. A variety of studies have used methods such as hedonic pricing, compensatory value and contingent valuation to assign a value to specific biodiversity assets. For example, the value of the urban forest in New York City alone has been calculated as over $5 billion (Nowak *et al.* 2002), nearby biodiversity can influence house prices, and residents are willing to pay millions of dollars to support urban biodiversity projects (e.g. Chen and Jim 2010).

Social values underpin the way people think about biodiversity

Another conception of values is used in other social disciplines such as psychology, sociology and anthropology. In these disciplines, values are seen as abstract conceptions of what people believe to be important, and act as guiding principles for the way people live their lives (Schwartz 1992). A broad range of values have been identified that can be mapped onto a two-dimensional space with one axis representing openness to change ←→ conservatism, and a second mapping on to self-enhancement (e.g. personal power, wealth) ←→ self-transcendence (e.g. altruism, social justice) (Schwartz 1992). A subset of people's values along the self-enhancement ←→ self-transcendence axis are thought to be particularly important in shaping people's responses to the environment and environmental concern: biospherism (e.g. protecting the environment), altruism (e.g. social justice) and egoism (e.g. wealth, power) (Stern 2000). Related conceptions of social value applied in an economic framework have identified a similar set of values: intrinsic values for nature (nature for its own sake), use values (e.g. production of timber) and non-use values (other human-centred values such as aesthetics, bequest for future generations) (Lockwood 1999).

Values, along with other mental constructs that influence the way we think about the world around us, are part of the cognitive hierarchy. In this hierarchy, values are usually placed at the bottom as they are considered relatively stable and influence other constructs higher up in the hierarchy. Beliefs and norms are both shaped by people's values (Ford *et al.* 2009). Beliefs are the things that people think are true. Social norms describe the rules that govern the behaviour of people in society – *what other people should do*. Personal norms are the rules that govern people's own behaviour – *what I should do*. Beliefs are an important component of people's response to wildlife (Manfredo *et al.* 2015). Beliefs that wildlife have their own rights, or that hunting is humane, have been shown to influence environmental attitudes. Beliefs about the consequences of management actions influence acceptability of biodiversity management (Stern 2000). Where people believe that a management action will adversely affect things they value, they tend to form negative judgements of that management action (Ford *et al.* 2009). Beliefs and norms are also relatively stable, although perhaps more amenable to change than values.

Beliefs are important drivers of conservation behaviour. The theory of planned behaviour shows that norms, attitudes and control beliefs (e.g. perceived ability to perform a behaviour, expected outcomes of behaviour) influence behavioural intentions, which in turn are related (sometimes weakly) to actual behaviours

(Ajzen 1985). Numerous studies have shown that the theory of planned behaviour is a useful predictor of environmental behaviours (Steg and Vlek 2009). An alternative theory of environmental behaviour based on moral concerns is the *Value–Belief–Norm* theory (Stern 2000) which shows that where people believe that environmental issues (e.g. pollution) will have negative consequences for things that they value (e.g. themselves, other people, biodiversity), they will be more likely to behave in ways that will reduce these consequences.

Attitudes are the judgements people make about the world around them (Heberlein 2012). Attitudes are typically seen as at the top of the cognitive hierarchy – people can have attitudes towards many things; they are influenced by constructs lower in the hierarchy (values, beliefs and norms) and they are more malleable than other cognitive constructs. A variety of attitudes related to biodiversity have been widely studied, such as the study of social acceptability of management, which has shown how beliefs and values shape people's attitudes to different kinds of biodiversity management, such as clearfell logging (Ford *et al.* 2009) and wildlife control (Zinn *et al.* 1998). There is likely to be an attitudinal component in the landscape preference judgements people make that is influenced by other cognitive constructs (Bjerke *et al.* 2006). Similarly, other environmental attitudes, such as the acceptability of management may influence preference judgements (Ford *et al.* 2012).

Managing biodiversity in line with perceptions and preferences

For urban residents, urban greenspaces are the dominant location of contact with the natural environment. These spaces should provide opportunities for enhancing quality of life, both through recreation and opportunities for contact with biodiverse nature, as well as conserving biodiversity. To do this, managers need to balance the provision of natural features with public facilities that provide opportunities for recreation and for the public to have contact with naturalistic environments (Voigt and Würster 2015). One of the challenges in managing urban greenspaces that provide both functions is to minimize damage from recreational activities (Qiu *et al.* 2013). Public parks, in particular those with remnant vegetation and high tree cover, are often areas where biodiversity can flourish (Shanahan *et al.* 2015). Shanahan and colleagues (2015) suggest that these parks with remnant vegetation can provide environmental and sustainability benefits as well as benefits for health and well-being.

Managing urban greenspaces in ways that support biodiversity conservation and human activity must also support a diverse range of people in the community with a range of values and preferences for nature interaction and recreation (Voigt and Würster 2015). As such, heterogeneity is likely to be an important factor in managing urban greenspaces that support the community. Different kinds of urban landscapes offer different levels of biodiversity, and can be managed for different aesthetic experiences (Table 8.1). Remnant forests are often used as a benchmark for urban biodiversity. They can contain many native species of plants and animals, and provide habitat for regionally threatened species. The composition and structure of remnant forests can be related to different theories of preference. For example, many forests have high levels of complexity, but relatively low levels of coherence or legibility. Forest understorey and infrastructure (e.g. paths) can be managed to increase mystery (e.g. winding paths) and coherence (delineating path edges) (Kaplan *et al.* 1998). In contrast, spontaneous grassland vegetation in wastelands can have high levels of diversity and support diverse invertebrate populations (Kowarik 2011), but provides a very different suite of psychological experiences – lower levels of complexity and refuge, but higher levels of 'prospect'. Designed meadows can have high levels of diversity but composition and management are carefully used to increase beauty (scenic aesthetic). The experiences provided by these biodiverse systems complement the experiences provided by traditional amenity landscapes. Neighbourhood parks and gardens may provide more preferred scenic aesthetic, but less opportunity to experience nature using an ecological aesthetic. And some landscapes provide other kinds of benefits outside the realm of preference – allotment/community gardens can provide food and opportunities for social interaction, while active sports fields provide opportunities for recreation that would not otherwise be available. Recognising the diverse opportunities available from different kinds of landscape facilitates a *portfolio of places* to be enjoyed by different people at different times, and for different reasons, so that people may choose landscapes based on their individual needs (Swanwick 2009). Biodiverse urban landscapes can play a complementary role in creating supportive and sustainable cities.

TABLE 8.1 Characteristics of different kinds of wild and non-wild landscapes, and how they influence the provision of aesthetic preferences. The levels of each characteristics potentially occurring in each landscape type are indicated by + low, ++ moderate +++ high.

Preference theories	Remnant (e.g. forest)	Vacant lot (e.g. grassland)	Designed meadow	Ornamental garden	Amenity park	Community/ allotment garden	Active sports field
Potential level of biodiversity	+++	+++	+++	+++	++	++	+
Information processing dimensions							
Complexity	+++	++	+++	+++	++	++	+
Mystery	++	+	+	++	++	+	+
Coherence	+	+	++	++	+++	+++	++
Legibility	+	+	++	++	+++	++	+
Aesthetic mode							
Ecological aesthetic	+++	+++	+++	++	++	++	+
Scenic aesthetic	+	+	++	+++	+++	++	++
Evolutionary dimensions							
Prospect	+	+++	+++	++	+++	++	+++
Refuge	+++	+	+	++	+++	+	+

Community engagement

Understanding the plurality of people's values, beliefs and attitudes that influence perceptions of biodiversity provides a useful pathway for meaningfully engaging with the community around biodiversity issues. Historically, engagement with the community on biodiversity and biodiversity management has focused on informing or educating the public about biodiversity, or the benefits for biodiversity of particular management actions (e.g. removing invasive species, introducing ecological disturbance such as fire). However, these engagements are not often successful. For example, research on conservation education programmes, run through community activity days, found that higher levels of awareness following the programme translated into greater interest in biodiversity. This did not, however, translate into reported changes in conservation or related behaviours over the longer term (Shwartz *et al.* 2012).

A greater understanding of the different ways that the community's values are expressed in relation to biodiversity provides an opportunity to engage with the broader community beyond those interested in biodiversity conservation, in ways that are meaningful to people with different values. A recent study exploring how people's values are expressed in relation to natural areas identified five groups of valued attributes of landscape (VALS) important to people: natural (e.g. diversity), cultural heritage, experiential (e.g. aesthetics, spirituality), social interaction and production (e.g. timber, food) (Kendal *et al.* 2015). Biodiversity is important to different people in different ways, and effective engagement needs to understand the diverse interests of the public. Understanding values provides a pathway to influence higher order cognitive constructs, and information that is aligned with values may be more effective in changing beliefs and attitudes. For example, ecological information can be used to change people's preference for biodiverse wetlands when this information is congruent with people's values (Straka *et al.* 2016).

Conclusion

This chapter highlighted the complementary role of social sciences in guiding biodiversity management decisions and practices. With increasing population and densification of cities around the world, urban greenspaces are likely to provide an important source of nature contact for city residents. But this raises questions about how to manage these spaces in ways that encourage nature interaction and recreation opportunities, as well as biodiversity conservation. It is important to understand that relationships between people and the environment are complex, with researchers only beginning to understand the role of biodiversity in influencing people's interactions with nature. New concepts such as 'biocultural diversity', the interrelationship between biodiversity and culture, are emerging in response to this complexity and criticisms of ecosystem services framework (Buizer *et al.* 2016).

It is likely perceptions and landscape preferences are important in shaping human responses to biodiversity. In particular, both landscape features (e.g. organization, cultural influences and context) and personal factors (e.g. environmental expertise, values, beliefs and norms) have an important role in in influencing perceptions, preferences and responses for biodiversity. Although approaches to managing urban greenspaces often advocate the provision of information on biodiversity, research suggests that this approach, in isolation, is unlikely to be effective in encouraging biodiversity appreciation or conservation for the broader public. Last, we highlight the importance of a varied approach to managing urban biodiversity as a way of supporting different members of the community and for maximizing urban biodiversity.

References

Ajzen I. (1985) From intentions to actions: A theory of planned behavior. In Kuhl J. and Beckmann J. (eds), *Action Control: From Cognition to Behavior*, pp. 11–39. Springer-Verlag, Berlin, Heidelberg, New York.

Appleton J. (1975) *The Experience of Landscape*. Wiley, London.

Bjerke T., Østdahl T., Thrane C. and Strumse E. (2006) Vegetation density of urban parks and perceived appropriateness for recreation. *Urban Forestry & Urban Greening* 5(1): 35–44.

Buizer M., Elands B. and Vierikko K. (2016) Governing cities reflexively: The biocultural diversity concept as an alternative to ecosystem services. *Environmental Science & Policy* 62: 7–13.

Bundy A.C., Lane S.J. and Murray E.A. (2002) *Sensory Integration—Theory and Practice*. F.A. Davis Company, Philadelphia, USA.

Chen W.Y. and Jim, C.Y. (2010) Resident motivations and willingness-to-pay for urban biodiversity conservation in Guangzhou (China). *Environmental Management* 45(5): 1052–1064.

Cilliers S. and Siebert S. (2011) Urban flora and vegetation: Patterns and processes. In J. Niemelä J.H., Breuste J.H., Guntenspergen G., McIntyre N.E., Elmqvist, T. and James P. (eds.), *Urban Ecology: Patterns, Processes, and Applications*. pp. 148–158, Oxford University Press, New York.

Dallimer M., Irvine K.N., Skinner A.M., Davies Z.G., Rouquette J.R., Maltby L.L., Warren P.H., Armworth P.R. and Gaston K.J. (2012) Biodiversity and the feel-good factor: Understanding associations between self-reported human well-being and species richness. *BioScience* 62(1): 47–55.

Dearborn D. and Kark S. (2009) Motivations for conserving urban biodiversity. *Conservation Biology* 24(2): 432–440.

de Groot W.T. and van den Born R.J.G. (2003) Visions of nature and landscape type preferences: an exploration in The Netherlands. *Landscape and Urban Planning* 63(3): 127–138.

Ford R.M., Williams K.J.H., Bishop I.D. and Webb T. (2009) A value basis for the social acceptability of clearfelling in Tasmania, Australia. *Landscape and Urban Planning* 90(3–4): 196–206.

Ford R.M., Williams K.J., Smith E.L. and Bishop I.D. (2012) Beauty, belief, and trust: Toward a model of psychological processes in public acceptance of forest management. *Environment and Behavior* 46: 476–506.

Fuller R.A., Irvine K.N., Devine-Wright P., Warren P.H. and Gaston K.J. (2007) Psychological benefits of greenspace increase with biodiversity. *Biology Letters* 3(4): 390–394.

Gobster P.P.H. (1999) An ecological aesthetic for forest landscape management. *Landscape Journal* 18(1): 54–64.

Gómez-Baggethun E. and Barton D.N. (2013) Classifying and valuing ecosystem services for urban planning. *Ecological Economics* 86: 235–245.

Heberlein T.A. (2012) Navigating environmental attitudes. *Conservation Biology* 26(4): 583–585.

Hope D., Gries C., Zhu W., Fagan W., Redman C., Grimm N., Nelson A.L., Martin C. and Kinzig A. (2003) Socioeconomics drive urban plant diversity. *Proceedings of the National Academy of Sciences of the United States of America* 100(15): 8788–8792.

Kaplan R. and Herbert E.J. (1987) Cultural and sub-cultural comparisons in preferences for natural settings. *Landscape and Urban Planning* 14: 281–293.

Kaplan R. and Kaplan S. (1989) *The Experience of Nature: A Psychological Perspective.* Cambridge University Press, Cambridge, UK.

Kaplan R., Kaplan S. and Ryan R. (1998) *With People in Mind: Design and Management of Everyday Nature.* Island Press, Washington, DC.

Kendal D., Williams K.J.H. and Williams N.S.G. (2012a) Plant traits link people's plant preferences to the composition of their gardens. *Landscape and Urban Planning* 105(1–2): 34–42.

Kendal D., Williams N.S.G. and Williams K.J.H. (2012b) Drivers of diversity and tree cover in gardens, parks and streetscapes in an Australian city. *Urban Forestry & Urban Greening* 11(3): 257–265.

Kendal D., Ford R.M., Anderson N.M. and Farrar A. (2015) The VALS: A new tool to measure people's general valued attributes of landscapes. *Journal of Environmental Management* 163: 224–233.

Kleiven J.O., Bjerke T. and Kaltenborn B.P. (2004) Factors influencing the social acceptability of large carnivore behaviours. *Biodiversity and Conservation* 13: 1647–1658.

Kowarik I. (2011) Novel urban ecosystems, biodiversity, and conservation. *Environmental Pollution* 159: 1974–1983.

Lee K.E., Williams K.J., Sargent L.D., Farrell C. and Williams N.S.W. (2014) Living roof preference is influenced by plant characteristics and diversity. *Landscape and Urban Planning* 122: 152–159.

Lockwood M. (1999) Humans valuing nature: Synthesising insights from philosophy, psychology and economics. *Environmental Values* 8(3): 381–401.

Manfredo M.J., Teel T.L. and Dietsch A.M. (2015) Implications of human value shift and persistence for biodiversity conservation. *Conservation Biology* 30(2): 287–296.

Nassauer J.I. (1995) Messy ecosystems, orderly frames. *Landscape Journal* 14(2): 161–170.

Nielsen A.B., van den Bosch M., Maruthaveeran S. and van den Bosch C.K. (2014) Species richness in urban parks and its drivers: A review of empirical evidence. *Urban Ecosystems* 17(1): 305–327.

Nowak D.J., Crane D.E. and Dwyer J.F. (2002) Compensatory value of urban trees in the United States. *Journal of Arboriculture* 28(4): 194–199.

Orians G.H. and Heerwagen J.H. (1992) Evolved responses to landscapes. In: Barkow J.H., Cosmides L., Tooby J. (eds), *The Adapted Mind: Evolutionary Psychology and the Generation of Culture*, pp 555–579. Oxford University Press, New York.

Qiu L., Lindberg S. and Nielsen A.B. (2013) Is biodiversity attractive? On-site perception of recreational and biodiversity values in urban green space. *Landscape and Urban Planning* 119: 136–146.

Schwartz S.H. (1992) Universals in the content and structure of values: Theoretical advances and empirical tests in 20 countries. In Zanna M.P. (ed.), *Advances in Experimental Social Psychology* 25: 1–65. Elsevier.

Shanahan D.F., Lin B.B., Gaston K.J., Bush R. and Fuller R.A. (2015) What is the role of trees and remnant vegetation in attracting people to urban parks? *Landscape Ecology* 30(1): 153–165.

Sheppard S.R. (2001) Beyond visual resource management: emerging theories of an ecological aesthetic and visible stewardship. *Forests and Landscapes: Linking Ecology, Sustainability and Aesthetics. IUFRO Research Series* 6: 149–172.

Shwartz A., Cosquer A., Jaillon A., Piron A., Julliard R., Raymond R., Simon L. and Prévot-Julliard A.C. (2012) Urban biodiversity, city-dwellers and conservation: how does an outdoor activity day affect the human-nature relationship? *PloS One* 7(6): e38642.

Shwartz A., Turbé A., Simon L. and Julliard R. (2014) Enhancing urban biodiversity and its influence on city-dwellers: An experiment. *Biological Conservation* 171: 82–90.

Steg L. and Vlek C. (2009) Encouraging pro-environmental behaviour: An integrative review and research agenda. *Journal of Environmental Psychology* 29(3): 309–317.

Stern P.C. (2000) Towards a coherent theory of environmentally significant behavior. *Journal of Social Issues* 56: 407–424.

Straka T.M., Kendal D. and van der Ree R. (2016) When ecological information meets high wildlife value orientations: Influencing preferences of nearby residents for urban wetlands. *Human Dimensions of Wildlife* 21(6): 538–554.

Swanwick C. (2009) Society's attitudes to and preferences for land and landscape. *Land Use Policy* 26: S62–S75.

van den Berg A.E., Koole S.L. and Cooper A.A. (2006) New wilderness in the Netherlands: An investigation of visual preferences for nature development landscapes. *Landscape and Urban Planning* 78(4): 362–372.

van den Berg A.E. and van Winsum-Westra M. (2010) Manicured, romantic, or wild? The relation between need for structure and preferences for garden styles. *Urban Forestry & Urban Greening* 9(3): 179–186.

Voigt A. and Würster D. (2015) Does diversity matter? The experience of urban nature's diversity: Case study and cultural concept. *Ecosystem Services* 12: 200–208.

Williams K.J. and Cary J. (2002) Landscape preferences, ecological quality, and biodiversity protection. *Environment and Behavior* 34(2): 257–274.

Yang B.E. and Kaplan R. (1990) The perception of landscape style: A cross-cultural comparison. *Landscape and Urban Planning* 19(3): 251–262.

Zinn H.C., Manfredo M.J., Vaske J.J. and Wittmann K. (1998) Using normative beliefs to determine the acceptability of wildlife management actions. *Society & Natural Resources* 11(7): 649–662.

9

BIODIVERSITY AND PSYCHOLOGICAL WELL-BEING

Kalevi Korpela, Tytti Pasanen and Eleanor Ratcliffe

Introduction

There is a growing body of evidence suggesting that contact with natural environments can promote well-being and psychological restoration from stress and/or fatigue, especially in the short term (Hartig *et al.* 2014). Much uncertainty still exists regarding the types of natural environments and the environmental qualities that promote these beneficial outcomes (Hartig *et al.* 2014), and the role of biodiversity especially is still unclear. Given that biodiversity has been widely recognised as one of the key features that support the global ecosystem and, consequently, humanity (Cardinale *et al.* 2012), better understanding of the relationship between human well-being and exposure to biodiverse environments is needed. To assess the current evidence on this topic, we searched for peer-reviewed original research articles and reviews from several scientific databases. In this illustrative review, we assess and summarise the main findings with an emphasis on psychological perspectives.

Defining biodiversity and well-being

Biodiversity and well-being are broad concepts including several definitions, outcomes and measures. *Biodiversity* includes species, genetic and ecosystem diversity (Convention on Biological Diversity, www.cbd.int/convention, accessed November 2016; Hammen and Settele 2011). Other definitions relate biodiversity to natural environments with greater or lesser visual or perceived complexity, or lushness or richness in species (e.g. Björk *et al.* 2008). Several studies use percent land area protected as a proxy for biodiversity and the proportion of threatened species and of highly disturbed land have been used to indicate decreased biodiversity (Lovell *et al.* 2014). Much of the literature we focus on in this chapter relates to species diversity and well-being.

Biodiversity is the foundation of ecosystem services; ecosystem services, in turn, contribute to human well-being (Millennium Ecosystem Assessment (MEA) 2005). Ecosystem services include *provisioning* services such as food, water and timber; *regulating* services such as the regulation of climate, disease and water quality; *cultural* services such as recreation, aesthetic enjoyment and spiritual fulfilment; and *supporting* services such as soil formation (MEA, 2005). European Environment Agency has recently developed the Common International Classification of Ecosystem Services (CICES, www.cices.eu, accessed November 2016). It includes three ecosystem services that, by definition, are the contributions that ecosystems make to human well-being: *Provisioning* services are all nutritional, material and energetic outputs from living systems. *Regulating and maintenance* cover all the ways in which living organisms can mediate or moderate the ambient environment that affects human performance. *Cultural* services cover all the non-material, and normally non-consumptive, outputs of ecosystems that affect physical and mental states of people. The MEA uses five dimensions to assess human *well-being* in relation to ecosystem services: basic material for a good life, freedom and choice, health, good social relations, and security. In Tzoulas *et al.*'s (2007) model, ecosystem services are linked to even more numerous aspects of health: ecosystem health, socio-economic, community, physical and psychological health (including relaxation from stress, positive emotions and cognitive capacity). In this chapter, our main focus is on individuals; thus, we will focus on emotional, cognitive and physiological aspects of well-being.

Echoing these different aspects, in psychology, research on well-being has tended to fall into two general groups (Ryan and Deci 2001). The *hedonic* viewpoint focuses on subjective well-being, frequently equated with happiness and defined as presence of positive affect (= feelings and emotions), low negative affect and greater life satisfaction. In contrast, the *eudaimonic* viewpoint focuses on psychological well-being, which is defined more broadly in terms of the fully functioning person. *Eudaimonic* well-being includes dimensions of autonomy, environmental mastery, personal growth, positive relations with others, purpose in life, self-acceptance, happiness together with meaningfulness, and self-actualisation (Ryan and Deci 2001). These positive aspects of functioning are thought to be promoted not only by social relationships and coping skills but also by exposure to environments that empower the person (Ryan and Deci 2001; Capaldi *et al.* 2015). Our aim in this chapter is to present studies examining the relationship of biodiversity to both *hedonic* and *eudaimonic* well-being.

In Lovell and colleagues' (2014) review of empirical studies focusing on the relationship between biodiversity and well-being, the conceptualisations of health and well-being ranged from life expectancy to sense of place or self-esteem. The outcome variables included infant mortality rate, incidence of low-weight babies, health behaviours such as physical activity, physical health status, psychological or emotional health, and community-level well-being. In relation to Tzoulas *et al.*'s (2007) model, outcomes representing psychological health (at the individual level) in particular are missing from these studies and we aim to contribute to this gap.

Different aspects of biodiversity affecting well-being

There seem to be at least three important aspects of the relationship between bio-diversity and well-being: the spatial scale of biodiversity; whether such biodiversity is perceived (subjectively rated) or actual (objectively measured), and evaluations/ effects of the presence or lack of biodiversity. These dimensions are all represented in the studies reviewed in this chapter, in which biodiversity has been operational-ised and examined as an independent variable.

The scale of biodiversity examined in relation to human well-being can vary considerably, from continents to specific environments including a certain land-mass, a specific nation-state, geographical regions in a country, and specific places (Lovell *et al.* 2014). Specific places are often most easily linked to individuals, especially in experimental research settings, and thus, in this chapter, our focus is mainly on the place and local level.

Concerning perceived vs. actual biodiversity there is mixed evidence regarding how well people are able to estimate species richness, i.e. species-level biodiver-sity (Fuller *et al.* 2007; Lindemann-Matthies *et al.* 2010; Dallimer *et al.* 2012). The diversity estimates of butterflies have been consistently inaccurate (Fuller *et al.* 2007; Dallimer *et al.* 2012), whereas some studies have found that lay people can approximately estimate the diversity of plants and even birds in an environment (Fuller *et al.* 2007; Lindemann-Matthies *et al.* 2010; Dallimer *et al.* 2012). Lee and Kendal (Chapter 8 of the current book) reviewed additional studies and expanded discussion on this issue. In the present chapter, both actual and perceived biodiver-sity in relation to well-being will be examined.

Presence of biodiversity or biodiversity gain can directly benefit health by ena-bling secure food production, preventing the spread of infectious diseases and providing nature-based medicine (Bernstein 2014). Increasing plant abundance can help to mitigate air pollution and thereby reduce incidences of respiratory and cardiovascular disease (Clark *et al.* 2014). On the other hand, biodiversity cannot always be judged as positive; allergies are one group of human diseases caused by the diversity of pollen, debris and proteins impacting the immune system (Hammen and Settele 2011). On the other hand, there are recent indications that exposure to microbial diversity, arising from the species richness of native flow-ering plants and land use types in the wider environment, is a potential pathway through which health benefits, such as lower prevalence of allergy, may arise (von Hertzen *et al.* 2011; Hanski *et al.* 2012). According to this 'biodiversity hypothesis', reduced contact of people with natural environmental features and biodiversity may adversely affect the human commensal microbiota and its immunomodulatory capacity and lead to immune dysfunction and chronic inflammatory disorder (von Hertzen *et al.* 2011; Hanski *et al.* 2012; Ruokolainen *et al.* 2015).

A dominant view seems to be that disturbance of ecosystems, and in particular biodiversity loss, may affect human health negatively through, for example, an increase in the spread of zoonotic diseases (Jones *et al.* 2008; Ostfeld 2009; Keesing *et al.* 2010) and declined access to food, clean water and raw materials (Sandifer *et al.* 2015). Changes in land use may reduce air and water quality which may increase the risk of respiratory problems and lung cancer (Wall *et al.* 2015). In a longitudinal study, a major change to the natural environment – the loss of 100 million trees to the emerald ash borer – increased county-level mortality related to cardiovascular and lower-respiratory-tract illness from 1990 to 2007 in 15 US States (Donovan *et al.* 2013).

Evidence of the links between biodiversity and well-being

Systematic reviews (e.g. Lovell *et al.* 2014) reveal mixed evidence for links between biodiversity and well-being, with some positive links and several null or even negative relationships, as well as inconsistency between subjective and objective measures of biodiversity. Lovell *et al.* (2014) found 17 primary research studies that were published in the past 12 years and primarily focused on relationships in Western developed countries. Nine studies showed one or more positive relationships between biodiversity and well-being. These benefits were manifested in a number of ways: from better mental health outcomes following exposure to associations with increased health-promoting behaviours. The relationships were most evident at a local scale, following immediate encounters or through presumed repeated exposures, and were found across the different study types and approaches. In all, Lovell *et al.* (2014) state that there is some evidence to suggest that biodiverse natural environments promote better health through exposure to pleasant environments or the encouragement of health-promoting behaviours. However, the overall evidence is inconclusive and fails to identify a specific role for biodiversity in the promotion of better health. This inconclusiveness seems to remain in those few studies – described in the following paragraphs – that we have found published after Lovell *et al.*'s (2014) review.

Mediating mechanisms between biodiversity and well-being

As implicated in Tzoulas *et al.*'s (2007) model, ecosystem services are linked to physical and psychological health. When explaining why physiological and psychological relaxation from stress, positive emotions and cognitive capacity might be affected by biodiversity we draw from restorative environments theories (see Hartig *et al.* 2014). We assume that processes described in those theories may be well-being outcomes as such, or act as mediating processes via which other well-being benefits emerge.

The theories include *Attention Restoration Theory* (ART; Kaplan and Kaplan 1989) and *Stress Reduction Theory* (SRT; Ulrich 1983). A consistent evidence for these complementary theories shows positive restoration effects after negative antecedent conditions, such as attentional fatigue or psychophysiological stress. Even passive short-term viewing of urban parks or woodlands (compared to built environments without natural elements) produces greater physiological changes toward relaxation, greater changes to positive emotions and faster recovery of attention-demanding cognitive performances (Hartig *et al.* 2014). Evidence of improved attentional functioning, emotional gains and lowered blood pressure and salivary cortisol has also been reported in field experiments using actual, 40–50-minute long walks in natural settings (Hartig *et al.* 2003; Park *et al.* 2010).

According to ART, restoration of attentional fatigue unfolds in place–person interactions that involve psychological distance from an individuals' usual routines ('Being Away'), effortless attention as drawn by objects in the environment, such as sunset, a fireplace, tree leaves ('Fascination'); immersion in a coherent physical or conceptual environment that is of sufficient scope to sustain exploration ('Coherence/extent') and a good match between personal inclinations and purposes, environmental supports for intended activities, and environmental demands for action ('Compatibility').

Stress Reduction theory assumes that perception of certain structural and depth properties of the visual array and general classes of environmental content (green plants, trees, water) rapidly evoke automatic positive affective (emotional) and physiological responses toward relaxation (Ulrich 1983; Hietanen *et al.* 2007). These responses may arise when complexity, i.e. the number of independently perceived elements in a scene, is moderate to high; the complexity has structural properties that establish a focal point and there is a moderate to high level of depth that can be perceived unambiguously; the ground surface texture tends to be homogeneous and even and is appraised as conducive to movement; a deflected vista (e.g. a curved path; a hill with trees) is present; threat is absent and water is present. Moreover, the majority of these elements are visual. However, none of these elements or preconditions of a preferred scene include biodiversity unless we speculate that complexity is positively related to biodiversity. Discussions regarding complexity occur in SRT and, to a certain extent, in ART within the concept of fascination and effortless attention through pattern (cf. Joye and van den Berg 2011). In studying links between biodiversity and psychological benefit, and especially restoration, the notion of environmental complexity also arises; i.e. more biodiverse environments may be more complex due to having a greater range of stimuli within them, for example via species richness. Such enhanced complexity may relate to preference for biodiverse environments (Carrus *et al.* 2015). We might also speculate that more biodiverse environments may be more likely to fascinate, to draw attention effortlessly due to richness of different kinds of stimuli. They also may create the sense of 'being away' more easily, again due to species richness that differs from everyday, urban environments. Looking from the personality perspective, for nature-related people (people with personality trait-like connectedness with nature), visiting more biodiverse environments might be conscious choices, compatible with their personal values and purposes, thus promoting well-being (Mayer *et al.* 2009). Some results of these processes are described in the following (psychological emotional outcomes), but due to lack of studies they mainly remain as speculations for future studies.

Other mechanisms indicated in the literature include a 'cultural pathway' as a mediating construct in the relationship between biodiversity and psychological benefits (Clark *et al.* 2014; Lovell *et al.* 2014). This pathway is related to cultural ecosystems services; i.e. the cultural values placed on nature through aesthetics, leisure and recreation (Church *et al.* 2011). It is assumed that people place cultural values on nature and that enhanced biodiversity can lead to 'cultural goods'. For example, higher wild species diversity and more aesthetically pleasing vistas encourage more frequent, longer or higher-quality experiences in nature, with resultant psychological benefits. On the other hand, loss in biodiversity may provide less opportunity to place cultural values on nature that can negatively affect human well-being. This implies that urban biodiversity per se may not be relevant, but rather its role as a marker of culturally defined environmental quality is. Another implication might be that biodiversity is effective only when it has cultural value. As there is a lack of studies on this pathway it remains outside of our focus in this chapter.

Biodiversity and cognitive human processes and outcomes

Biodiversity is related to self-reported cognitive processing of environments, as well as cognitive outcomes in such environments. With regard to cognitive appraisals or processing, a survey study investigated the effects of objectively assessed biodiversity in six urban sites and self-reported nature vs. urban relatedness on perception of the lushness of urban green space (Gunnarsson *et al.* 2016). Participant perceptions of lushness and sound-related importance of urban greenery was highest where urban biodiversity was highest. Self-reports of the importance of trees and plants for perception of bird species in urban greenery followed the same pattern, although differing only between high and medium/low biodiversity conditions. Nature and urban relatedness moderated these perceptions (see section on individual and contextual differences). Self-reported cognitive processes have also been identified as potential mediating mechanisms between biodiversity and affective (emotional) wellbeing outcomes. For example, positive relationships between perceived bird biodiversity and positive affect after an outdoor walk were mediated by the perceived restorativeness of environment including fascination and sense of "being away" from everyday concerns (Marselle *et al.* 2016).

Evidence regarding links between biodiversity and cognitive outcomes is largely limited to qualitative research and to that which encompasses self-reported psychological restoration, of which cognition is one component. For example, in interviews with wildlife tourists, participants reported positive cognitive outcomes such as flow, or being completely absorbed in the moment, from observing a diverse range of animals, as well as feeling like time had stopped (Curtin 2009). They also reported opportunities to think, reflect, and contemplate. In a study of urban green space users, Fuller *et al.* (2007) identified objectively measured plant richness, bird (but not butterfly) richness, and number of habitat types as significant, positive predictors of reflection, a cognitive aspect of subjective wellbeing, whereas Dallimer *et al.* (2012) found perceived, but not actual, biodiversity of plants, butterflies and birds to be predictive of reflection.

Biodiversity and physiological human processes and outcomes

Although a number of studies have shown a connection between greater biodiversity and better physiological health (Sandifer *et al.* 2015), the only relationship where causality has been reliably detected suggests that exposure to environmental biodiversity improves the maintenance of healthy immune systems and reduces the prevalence of inflammatory-based diseases (Hanski *et al.* 2012; Rook 2013; Bernstein 2014; Hough 2014). However, in a comprehensive UK study, bird species richness as an indicator of local biodiversity was associated with good health (but not with bad health) prevalence measured by self-reported general health status (Wheeler *et al.* 2015). We found only one study on direct physiological outcomes where the level of diversity was manipulated and in that experiment, the results were inconsistent (Cracknell *et al.* 2016).

Regarding individuals' health behaviour and outcomes, Tilt *et al.* (2007) studied subjective and objective measures of greenness (which have shown a positive correlation with biodiversity) in relation to self-reported physical activity and the self-reported body mass index (BMI). The number of monthly walking trips was greater among those who had identified more nature-related objects within 800 metres from their home (*subjective greenness*) whereas no relationship to the normalised difference vegetation index (*objective greenness*) was found. BMI was lower in areas that were objectively greener and provided good accessibility to services, compared with areas with good accessibility but low levels of objective greenness. A study of suburban and rural residents observed that environmental vegetation lushness, taken as a proxy for biodiversity, was significantly and positively associated with time spent on physical activities, but not self-reported health or vitality (Björk *et al.* 2008). In a cross-sectional study, vegetation complexity (objective measure) was not related to either physical activity or blood pressure (Shanahan *et al.* 2016).

Biodiversity and psychological/emotional processes and outcomes

Feelings, emotions and arousal

The majority of evidence for links between biodiversity and psychological well-being involves affective (=emotional) processes and outcomes. This includes both qualitative and self-reported quantitative measures of affect. For example, in her qualitative research involving wildlife tourists, Curtin (2009) observed that participants expressed strong positive affect as a result of watching a range of animals (both on tourist trips and in the context of their daily life); in some cases this was linked to biodiversity, e.g. 'I am quite happy' or 'It is the sheer diversity . . .' (p. 457). However, acknowledgement of the negative valence of some aspects of animal behaviour was noted, such as for inter-animal predation. Participants also expressed how arousing or intense the encounters were, in a positive context.

With regard to quantitative research, Barton *et al.* (2009) conducted a survey for visitors in areas with high natural and heritage value, with half the sample interviewed before entering the site and the other half as they were leaving the site. Those leaving the areas reported less anger, depression and confusion, more vigour, and slightly better self-esteem compared with the ones interviewed before entering the area.

Evidence at a species level regarding links between biodiversity and affective well-being is mixed, both in terms of the direction of relationships and the type of biodiversity involved. Cracknell *et al.* (2015) observed better mood and decreased arousal in response to viewing increased diversity of fish in a public aquarium, in addition to the physiological outcomes described in the previous section. Similar positive results were found in the study by Fuller *et al.* (2007), discussed in the section on cognitive processes, where objectively measured plant and bird species richness, as well as the number of habitat types, were significant positive predictors of affective aspects of subjective well-being such as place identity and attachment and continuity with the past. In contrast, Dallimer *et al.* (2012) found perceived, but not actual, biodiversity of plants, butterflies and birds to be predictive of these affective aspects of subjective psychological well-being.

Contrary to the findings above, Marselle *et al.* (2015) reported that perceived bird biodiversity was positively related to negative affect after a walk. Suggested explanations focus on the type of birds; e.g. that, as more birds were heard, so too were birds with unpleasant sounds/calls. Other types of perceived biodiversity (e.g. butterflies, plants, trees) were not predictive of emotional well-being after the walk. In a later study, Marselle *et al.* (2016) observed that perceived bird, butterfly and tree biodiversity properties were indirect positive predictors of self-reported affect after a walk, mediated via perceived restorativeness; this is discussed further in the following section on restoration.

The studies above concern short-term affective response. However, studies concerning longer-term affective well-being have found either no link between exposure to more biodiverse urban environments and mental health (Shanahan *et al.* 2016) or have even found negative links (Huby *et al.* 2006).

Restoration

Several studies on links between biodiversity and well-being focus on the ability of biodiverse environments to engender cognitive and affective change, primarily in the form of perceived restoration from stress and/or mental fatigue. For example, Curtin's (2009) participants reflected on the ability of their wildlife encounters to generate restorative benefits. They were referred to as 'uplifting' or otherwise able to counter negative affective states. In some cases these psychological changes were linked with spirituality or opportunities for reflection on something greater than oneself.

As noted in the previous section, Marselle *et al.* (2016) found that cognitive and affective aspects of perceived restorativeness, in the form of fascination, being away and compatibility, mediated relationships between bird biodiversity and post-walk positive affect, happiness and negative affect. Marselle *et al.* (2016) interpret these findings as evidence for restorativeness as a key mechanism in links between biodiversity and well-being, in that increased perceptions of bird biodiversity may lead to greater perceptions of an environment's restorative value, and thereby increase affective benefits achieved *in situ*.

In a study involving users of five urban and peri-urban outdoor environments in Italy, Scopelliti *et al.* (2012) observed that more biodiverse environments were rated as more restorative and more psychologically and physically beneficial than less biodiverse environments. Similarly, Carrus *et al.* (2015) reported that, compared to environments low in biodiversity, high-biodiversity environments were rated as more restorative and psychologically beneficial. This effect was greatest in environments that were urban rather than peri-urban. Links between increased biodiversity and psychological benefits were mediated by the perceptions of fascination, being away, coherence and compatibility. In explaining these findings, Carrus *et al.* (2015) suggest that enhanced biodiversity may compensate for negative effects of urbanism on perceived psychological outcomes. It is notable that, in these two studies, biodiversity was not operationalised explicitly as a quantitative difference in species richness, but rather as qualitative distinctions made by the study authors based on parameters used in forestry science that include species richness (cf. Carrus *et al.* 2015).

Eudaimonic well-being

Links between biodiversity and eudaimonic aspects of well-being, such as life satisfaction or community-based well-being, appear inconclusive. Some studies have reported no relationship between biodiversity and social well-being (Shanahan *et al.* 2016), while others have provided mixed evidence. For instance, in a study of residents of Australian urban areas, Luck *et al.* (2011) assessed relationships between biodiversity and personal well-being (life satisfaction), neighbourhood well-being (satisfaction with local area), and personal connection to nature. Neighbourhood well-being was most strongly and positively predicted by bird species richness and vegetation cover, while personal well-being and connection to nature were more weakly predicted by these biodiversity variables. In all cases demographic factors (e.g. age, gender and education) were strong and more consistent predictors of well-being.

Björk *et al.* (2008) and de Jong *et al.* (2012) comment on inconsistencies in the literature regarding how well-being is measured, which is especially relevant given that these studies focus on an eudaimonic perspective – i.e. life satisfaction as well-being, rather than hedonic evaluations of affect. In particular, there is heterogeneity in such measures of eudaimonic well-being, with Björk *et al.* (2008) using a mix of single- and multiple-item scales, and Luck *et al.* (2011) using only the latter. As such, comparison of findings is limited.

Individual and contextual differences in the relationship between biodiversity and well-being

Studies on individual differences are few and provide only anecdotal evidence. There is need for greater consideration of contextual and demographic factors when considering links between biodiversity and well-being (Björk *et al.* 2008; Luck *et al.* 2011). Björk *et al.* (2008) observed that biodiversity in the form of vegetation lushness was associated with neighbourhood satisfaction for tenants but not homeowners. In a second study of Swedish suburban and rural residents, de Jong *et al.* (2012) consolidated self-reported environmental quality with objective measures of the same, producing a *Scania Green Score* (SGS). SGS was positively related to self-reported neighbourhood satisfaction, physical activity and health. More specifically, lush environments were positively associated with self-reported physical activity and neighbourhood satisfaction. In the case of the latter relationship, this was apparent only among participants living 'in a flat or student room' (p. 1377). Thus, it seems that lush environments may lead to greater neighbourhood satisfaction for those living in smaller or possibly lower-quality built environments so that the lushness of the environment perhaps compensates for built-environment limitations.

Pereira *et al.* (2005) conducted interviews with residents of a rural and primarily agricultural community in Portugal. Participants did not explicitly connect biodiversity with their self-reported well-being, but did mention the role of biodiversity in supporting environmental quality. Biodiversity was connected to both positive and negative evaluations; it was related to aesthetic and positively valued appraisals, but also negative affect as a result of crop damage, livelihood loss through wildlife predation and resulting negative effects on livelihood (and, by extension, well-being). Pereira *et al.* (2005) point out that agriculture is dependent on intentional, human-directed control of biodiversity, and that negative evaluations of biodiversity or unclear links with well-being may have been a function of this specific study population.

In a study on biodiversity and perceptions of urban green space, people rating themselves as highly nature related were shown to prefer urban green space aesthetics and to value greenery-related sounds more, and to attach greater importance to trees and plants in their perception of bird species, than less nature related persons (Gunnarsson *et al.* 2017). Furthermore, a study in Australia revealed that nature relatedness was a much stronger determinant of visitation to a nearby park than the amount of nearby available park area (Lin *et al.* 2017). People with high nature relatedness typically had greater vegetation cover around their home as well as spent more time within these yards (Lin *et al.* 2017). Moreover, tree cover and remnant vegetation cover (naturally occurring or rehabilitated vegetation with high ecological value) attracted people to parks only weakly except for nature related persons (Shanahan *et al.* 2015b). These results point to an interesting but yet poorly understood role for personal nature relatedness in health and well-being benefits from nature (see Capaldi *et al.* 2015).

Overall, studies examining contextual or individual differences in links between biodiversity and well-being are few, but this number appears to be growing in recent years. The studies that do exist identify that such links may differ across cultural and socio-demographic contexts.

Gaps in the literature and suggestions for future studies

We found very little theoretical work on the psychological mechanisms and processes linking urban biodiversity and well-being. Studies linking exposure to green space and blue space to well-being abound, but the investigation of the specific role of biodiversity in this relationship is in the early phase. The question of why gain or loss of biodiverse environments might influence, for example, positive affect, self-esteem or mental fatigue remain largely unanswered. Findings among the studies that we selected for the current, illustrative rather than comprehensive review, were mixed with positive, negative or no relationship found. Measures of biodiversity ranged from objective to perceived proxy measures. It is also noteworthy that many of the studies we have cited investigated animal species diversity, particularly birds, rather than plant species diversity. Furthermore, the majority of the studies focused on hedonic well-being (positive affect, pleasantness), particularly affect and arousal, with less evidence on eudaimonic well-being (self-actualisation, meaningful life). Very few experimental studies that manipulate the levels of biodiversity in the context of well-being outcomes seem to exist. Longitudinal evidence on biodiversity and health is currently limited to physiological well-being/mortality. More studies assessing causalities on the psychological perspective are needed, as has also been noted by other review studies (e.g. Sandifer *et al.* 2015; Shanahan *et al.* 2015a). There is very little knowledge of the types and qualities of environments and ecological features that affect well-being. Furthermore, the relative contributions of different factors, such as biodiversity, proximity or accessibility, are not known (Keniger *et al.* 2013). We have adapted a framework for future studies (Figure 9.1), based on the framework outlined by Shanahan and colleagues (2015a; 2016). Our examples on mediators, effects on people, and health benefits are based on reviews by Kuo (2015) and Hartig *et al.* (2014). Examples for moderators come from findings on gender by Annerstedt *et al.* (2012), on age by Astell-Burt *et al.* (2014), on income by Mitchell and Popham (2008) and on nature relatedness by Lin *et al.* (2017).

Studies that examine differences in links between urban biodiversity and well-being between individuals and cultures are few, but growing. Such studies are also urgently required in wider research on well-being and nature per se, in order to better understand not just whether, but for whom, these environments can offer psychological benefits. The literature examined in this chapter provides insights into the plurality of natural settings that can affect physiological and psychological states, but much research remains to be done to provide a convincing case for a positive role of urban biodiversity in such phenomena. In these efforts, multidisciplinary co-operation is needed to combine expertise on biodiversity, mental and physical

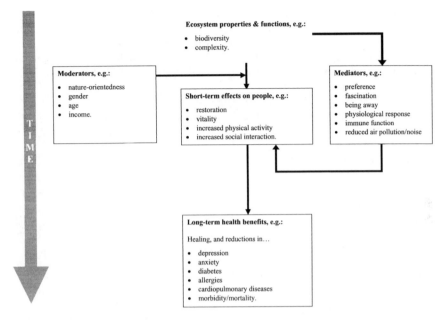

FIGURE 9.1 Proposed framework for future studies of interactions between urban ecosystem properties, such as biodiversity, and well-being including examples of potential, not proven, moderators, mediators and effects. Adapted from Shanahan *et al.* (2015a), Kuo (2015), and Hartig *et al.* (2014).

health. Obviously, not only ecology, biology and psychology but also landscape architecture, urban planning, forestry and medicine, to name just a few, have to come together in a combined effort to conduct qualified research and summarise it in order to get a comprehensive understanding of the relationship between bio-diversity and human well-being (cf. Tzoulas *et al.* 2007). Methodologies that can convey data and results from behavioural, social, biological and medical sciences for environmental planners, designers and managers in a comprehensible format are needed. One such example is the softGIS methodology which is a collection of internet-based surveys that allow the place-based study of human experiences (Kahila and Kyttä 2009). In practice, it seems crucial to integrate access to nature and urban planning with the health care sector and services to support prevention of illness and promotion of health (Africa *et al.* 2014).

References

Africa J., Logan A., Mitchell R., Korpela K., Allen D., Tyrväinen L., Nisbet E., Li Q., Tsunetsugu Y., Miyazaki Y. and Spengler J.; on behalf of the NEI Working Group. (2014) *The Natural Environments Initiative: Illustrative Review and Workshop Statement*. Center for Health and the Global Environment at the Harvard School of Public Health. Available at www.chgeharvard.org/sites/default/files/resources/Paper-NaturalEnvironmentsInitiative_0.pdf (accessed 5 August 2017).

Annerstedt M., Ostergren P.-O., Björk J., Grahn P., Skärbäck E. and Währborg P. (2012) Green qualities in the neighbourhood and mental health: Results from a longitudinal cohort study in Southern Sweden. *BMC Public Health* 12(1): 337.

Astell-Burt T., Mitchell R. and Hartig T. (2014) The association between green space and mental health varies across the lifecourse: A longitudinal study. *Journal of Epidemiology and Community Health* 68: 578–583.

Barton J., Hine R. and Pretty J. (2009) The health benefits of walking in greenspaces of high natural and heritage value. *Journal of Integrative Environmental Sciences* 6: 261–278.

Bernstein A.S. (2014) Biological diversity and public health. *Annual Review of Public Health* 35(1): 153–167.

Björk J., Albin M., Grahn P., Jacobsson H., Ardö J., Wadbro J., Östergren P.-O. and Skärbäck E. (2008) Recreational values of the natural environment in relation to neighbourhood satisfaction, physical activity, obesity and wellbeing. *Journal of Epidemiology & Community Health* 62: 1–7.

Capaldi C.A., Passmore H.-A., Nisbet E.K., Zelenski J.M. and Dopko R.L. (2015) Flourishing in nature: A review of the benefits of connecting with nature and its application as a wellbeing intervention. *International Journal of Wellbeing* 5(4): 1–16.

Cardinale B.J., Duffy J.E., Gonzalez A., Hooper D.U., Perrings C., Venail P., Narwani A., Mace G.M., Tilman D., Wardle D.A., Kinzig A.P., Daily G.C., Loreau M., Grace J.B., Larigauderie A., Srivastava D.S. and Naeem S. (2012) Biodiversity loss and its impact on humanity. *Nature* 486(7401): 59–67.

Carrus G., Scopelliti M., Colangelo G., Ferrini F., Salbitano F., Agrimi M., Portoghesi L., Semenzato P., Sanesi G., Lafortezza R. and Colangelo G. (2015) Go greener, feel better? The positive effects of biodiversity on the well-being of individuals visiting urban and peri-urban green areas. *Landscape & Urban Planning* 134: 221–228.

Church A., Burgess J. and Ravenscroft N. (2011) *Cultural Services.* The UK National Ecosystem Assessment Technical Report (pp. 633–692). UNEP-WCMC, Cambridge.

Clark N.E., Lovell R., Wheeler B.W., Higgins S.L., Depledge M.H. and Norris K. (2014) Biodiversity, cultural pathways, and human health: A framework. *Trends in Ecology & Evolution* 29: 198–204.

Cracknell D., White M.P., Pahl S., Nichols W.J. and Depledge M.H. (2016) Marine biota and psychological well-being: A preliminary examination of dose–response effects in an aquarium setting. *Environment & Behavior* 48: 1242–1269.

Curtin S. (2009) Wildlife tourism: The intangible, psychological benefits of human–wildlife encounters. *Current Issues in Tourism* 12: 451–474.

Dallimer M., Irvine K.N., Skinner A.M., Davies Z.G., Rouquette J.R., Maltby L.L., Warren P.H., Armsworth P.R. and Gaston K.J. (2012) Biodiversity and the feel-good factor: Understanding associations between self-reported human well-being and species richness. *Bioscience* 62(1): 47–55.

de Jong K., Albin M., Skärbäck E., Grahn P. and Björk J. (2012) Perceived green qualities were associated with neighborhood satisfaction, physical activity, and general health: Results from a cross-sectional study in suburban and rural Scania, southern Sweden. *Health & Place* 18: 1374–1380.

Donovan G.H., Butry D.T., Michael Y.L., Prestemon J.P., Liebhold A.M., Gatziolis D. and Mao M.Y. (2013) The relationship between trees and human health: Evidence from the spread of the emerald ash borer. *American Journal of Preventive Medicine* 44(2): 139–145.

Fuller R.A., Irvine K.N., Devine-Wright P., Warren P.H. and Gaston K.J. (2007) Psychological benefits of greenspace increase with biodiversity. *Biology Letters* 3(4): 390–394.

Gunnarsson B., Knez I., Hedblom M. and Ode Sang Å. (2017) Effects of biodiversity and environment-related attitude on perception of urban green space. *Urban Ecosystems* 20: 37–49.

Hammen V.C. and Settele J. (2011) Biodiversity and the loss of biodiversity affecting human health, in Nriagu J.O. (ed.), *Encyclopedia of Environmental Health* (pp. 353–362). Elsevier, Burlington, VT.

Hanski I., von Hertzen L., Fyhrquist N., Koskinen K., Torppa K., Laatikainen T., Karisola P., Auvinen P., Paulin L., Mäkelä M.J., Vartiainen E., Kosunen T.U., Alenius H. and Haahtela T. (2012) Environmental biodiversity, human microbiota, and allergy are interrelated. *Proceedings of the National Academy of Sciences of the United States of America* 109: 8334–8339.

Hartig T., Evans G.W., Jamner L.D., Davis D.S. and Gärling T., (2003) Tracking restoration in natural and urban field settings, *Journal of Environmental Psychology* 23(2): 109–123.

Hartig T., Mitchell R., De Vries S. and Frumkin H. (2014) Nature and health. *Annual Review of Public Health* 35: 207–228.

Hietanen J.K., Klemettilä T., Kettunen J.E. and Korpela K.M. (2007) What is a nice smile like that doing in a place like this? Automatic affective responses to environments influence the recognition of facial expressions. *Psychological Research* 71: 539–552.

Hough R.L. (2014) Biodiversity and human health: Evidence for causality? *Biodiversity and Conservation* 23: 267–288.

Huby M., Cinderby S., Crowe A.M., Gillings S., McClean C.J., Moran D., Owen A. and White P.C.L. (2006) The association of natural, social and economic factors with bird species richness in rural England. *Journal of Agricultural Economics* 57(2): 295–312.

Jones K.E., Patel N.G., Levy M.A., Storeygard A., Balk D., Gittleman J.L. and Daszak P. (2008) Global trends in emerging infectious diseases. *Nature* 451(7181): 990–993.

Joye Y. and van den Berg A. (2011) Is love for green in our genes? A critical analysis of evolutionary assumptions in restorative environments research. *Urban Forestry & Urban Greening* 10: 261–268.

Kahila M. and Kyttä M. (2009) SoftGIS as a bridge builder in collaborative urban planning, in Geertman, S. and Stillwell, J. (eds) *Planning Support Systems: Best Practices and New Methods* (pp. 389–411). Springer, Dordrecht.

Kaplan R. and Kaplan S. (1989) *The Experience of Nature: A Psychological Perspective.* Cambridge University Press, New York.

Keesing F., Belden L.K., Daszak P., Dobson A., Harvell C.D., Holt R.D., Hudson P., Jolles A., Jones K.E., Mitchell C.E., Myers S.S., Bogich T. and Ostfeld R.S. (2010) Impacts of biodiversity on the emergence and transmission of infectious diseases. *Nature* 468: 647–652.

Keniger L., Gaston K.J., Irvine K. and Fuller R. (2013) What are the benefits of interacting with nature? *International Journal of Environmental Research and Public Health* 10(3): 913–935.

Kuo M. (2015) How might contact with nature promote human health? Promising mechanisms and a possible central pathway. *Frontiers in Psychology* 6: 1093.

Lin B.B., Gaston K.J., Fuller R.A., Wu D., Bush R. and Shanahan D.F. (2017) How green is your garden?: Urban form and socio-demographic factors influence yard vegetation, visitation, and ecosystem service benefits. *Landscape and Urban Planning* 157: 239–246.

Lindemann-Matthies P., Junge X. and Matthies D. (2010) The influence of plant diversity on people's perception and aesthetic appreciation of grassland vegetation. *Biological Conservation* 143(1): 195–202.

Lovell R., Wheeler B.W., Higgins S.L., Irvine K.N. and Depledge M.H. (2014) A systematic review of the health and well-being benefits of biodiverse environments. *Journal of Toxicology & Environmental Health*, Part B 17: 1–20.

Luck G.W., Davidson P., Boxall D. and Smallbone L. (2011) Relations between urban bird and plant communities and human well-being and connection to nature. *Conservation Biology* 25: 816–826.

Marselle M.R., Irvine K.N., Lorenzo-Arribas A. and Warber S.L. (2015) Moving beyond green: Exploring the relationship of environment type and indicators of perceived environmental quality on emotional well-being following group walks. *International Journal of Environmental Research and Public Health* 12: 106–130.

Marselle M.R., Irvine K.N., Lorenzo-Arribas A. and Warber S.L. (2016) Does perceived restorativeness mediate the effects of perceived biodiversity and perceived naturalness on emotional well-being following group walks in nature? *Journal of Environmental Psychology* 46: 217–232.

Mayer F.S., Frantz C.M., Bruehlman-Senecal E. and Dolliver K. (2009) Why is nature beneficial?: The role of connectedness to nature. *Environment and Behavior* 41: 607–643.

Millennium Ecosystem Assessment (2005) *Ecosystems and Human Well-being: Biodiversity Synthesis*. World Resources Institute, Washington, DC. Available at www.millenniumassessment. org/documents/document.354.aspx.pdf (accessed 15 September 2016).

Mitchell R. and Popham F. (2008) Effect of exposure to natural environment on health inequalities: An observational population study. *The Lancet* 372, 1655–1660.

Park B.-J., Tsunetsugu Y., Kasetani T., Kagawa T. and Miyazaki Y. (2010) The physiological effects of Shinrin-yoku (taking in the forest atmosphere or forest bathing): Evidence from field experiments in 24 forests across Japan. *Environmental Health and Preventive Medicine* 15: 18–26.

Ostfeld R. S. (2009) Biodiversity loss and the rise of zoonotic pathogens. *Clinical Microbiology and Infection* 15: 40–43.

Pereira E., Queiroz C., Pereira H. and Vicente L. (2005) Ecosystem services and human well-being: A participatory study in a mountain community in Portugal. *Ecology and Society* 10: 1–23.

Rook G.A. (2013) Regulation of the immune system by biodiversity from the natural environment: An ecosystem service essential to health. *Proceedings of the National Academy of Sciences of the United States of America* 110(46): 18360–18367.

Ruokolainen L., von Hertzen L., Fyhrquist N., Laatikainen T., Lehtomäki J., Auvinen P., Karvonen A.M., Hyvärinen A., Tillmann V., Niemelä O., Knip M., Haahtela T., Pekkanen J. and Hanski I. (2015) Green areas around homes reduce atopic sensitization in children. *Allergy* 70: 195–202.

Ryan R.M. and Deci E.L. (2001) On happiness and human potentials: A review of research on hedonic and eudaimonic well-being. *Annual Review of Psychology* 52: 141–166.

Sandifer P.A., Sutton-Grier A.E. and Ward B.P. (2015) Exploring connections among nature, biodiversity, ecosystem services, and human health and well-being: Opportunities to enhance health and biodiversity conservation. *Ecosystem Services* 12: 1–15.

Scopelliti M., Carrus G., Cini F., Mastandrea S., Ferrini F., Lafortezza R., Agrimi M., Salbitano F., Sanesi G. and Semenzato P. (2012) Biodiversity, perceived restorativeness and benefits of nature: A study on the psychological processes and outcomes of on-site experiences in urban and peri-urban green areas in Italy. In Kabisch S., Kunath A., Schweizer-Ries P. and Steinführer A. (eds), *Vulnerability, Risks and Complexity: Impacts of Global Change on Human Habitats* (pp. 255–270). Hogrefe, Göttingen.

Shanahan D.F., Bush R., Gaston K.J., Lin B.B., Dean J., Barber E. and Fuller R.A. (2016) Health benefits from nature experiences depend on dose. *Scientific Reports* 6: 28551.

Shanahan D.F., Lin B.B., Bush R., Gaston K.J., Dean J.H., Barber E. and Fuller R.A. (2015a) Toward improved public health outcomes from urban nature. *American Journal of Public Health* 105: 470–477.

Shanahan D.F., Lin B.B., Gaston K.J., Bush R. and Fuller R.A. (2015b) What is the role of trees and remnant vegetation in attracting people to urban parks? *Landscape Ecology* 30: 153–165.

Tilt J. H., Unfried T.M. and Roca B. (2007) Using objective and subjective measures of neighborhood greenness and accessible destinations for understanding walking trips and BMI in Seattle, Washington. *American Journal of Health Promotion* 21: 371–379.

Tzoulas K., Korpela K., Venn S., Yli-Pelkonen V., Kazmierczak A., Niemelä J. and James P. (2007) Promoting ecosystem and human health in urban areas using green infrastructure: A literature review. *Landscape and Urban Planning* 81: 167–178.

Ulrich R.S. (1983) Aesthetic and affective response to natural environment. In Altman I. and Wohlwill J.F. (eds), *Human Behavior and Environment. Vol. 6, Behavior and the Natural Environment* (pp. 85–125). Plenum Press, New York.

von Hertzen L., Hanski I. and Haahtela T. (2011) Natural immunity: Biodiversity loss and inflammatory diseases are two global megatrends that might be related. *EMBO Rep* 12: 1089–1093.

Wall D.H., Nielsen U.N. and Six J. (2015) Soil biodiversity and human health. *Nature* 528(7580): 69–76.

Wheeler B.W., Lovell R., Higgins S.L., White M.P., Alcock I., Osborne N.J., Husk K., Sabel C.E. and Depledge M.H. (2015) Beyond greenspace: An ecological study of population general health and indicators of natural environment type and quality. *International Journal of Health Geographics* 14(1): 17.

10

GOVERNANCE PERSPECTIVES ON URBAN BIODIVERSITY

Sara Borgström

Introduction

Nature is not isolated from society, but rather an expression of a co-evolution of social and ecological processes within social–ecological systems (Folke 2006). Therefore, any attempt to understand nature requires the inclusion of i) social aspects, such as actors, power, history, planning and management, ii) ecological aspects, such as biodiversity patterns, hydrology and nutrient cycling, iii) and the interactions between these, for example land use change (Berkes and Folke 1998). In cities, humans are the driving forces in the system due to the density of settlements where the landscape is shaped primarily on the basis of human needs. Therefore, nature and biodiversity in cities are to a large degree determined by social factors (Grimm *et al.* 2008). Because many people need to share limited space in cities, a well-developed governance system is necessary for securing decent living conditions for the population. For example, urban land use is often strictly directed by a number of formal institutions, i.e. socially-based constraints that shape human choice, behaviour and interaction in the society – *the rules of the game* (North 1991). These institutions, e.g. land use plans, are of importance also for urban biodiversity.

This chapter discusses urban biodiversity conservation from a governance perspective with particular focus on formal institutions such as legal frameworks and regulations, and how these are expressed in the process of urban development. From a presentation of the main elements of governance, the chapter focuses on two approaches related to urban biodiversity conservation: (1) the conventional nature conservation framework of establishing protected areas, and (2) the ecosystem services framework for integrating nature into urban planning processes. The chapter further elaborates on the challenges of these two approaches to urban biodiversity conservation as well as emerging ways forward.

Elements of governance

The concept of *governance* has been developed within social science and depicts the broadening of the understanding of societal decision making, where governance includes not only hierarchical political steering as in government, but also network forms of collaborative decision making (Kooiman 2003). This implies an inclusion of a wider range of actors in decision making by adding private and civic society actors to the process that has traditionally been restricted to the public sector. These private and civic society actors come with a diversity of preconditions for engaging in the process such as different resources, agendas and power (Kooiman 2003; Peters and Pierre 2004). A diversification of actors also implies that the representative decision-making process is complemented with means of deliberation such as participatory planning and co-management (Fung and Wright 2001; Carlsson and Berkes 2005). The actors in a governance system interact both horizontally and vertically, and, therefore, the governance levels are interdependent, as captured by the concept of *multilevel governance* (Peters and Pierre 2004).

Institutions, organisations and policies

Governance is inherently complex and there are many suggestions of how to disentangle the interrelated compartments and processes. Lange *et al.* (2013) divide governance into political processes (*politics*), institutional structures (*polity*) and policy content (*policy*). This is a useful approach to identify and understand governance around a certain issue, such as urban biodiversity conservation. In this chapter, the focus is placed on institutional structures. These socially constructed institutions include both formal rules, such as written laws and contracts, and informal rules, such as unwritten codes of social life, e.g. norms, customs and taboos. Furthermore, institutions include most types of organisations, from governmental agencies to families and communities, since those are often defined in terms of rules and norms (Imperial 1999). Organisations can be said to be the arena where institutions are played out. The operationalisation of institutions often takes the form of policies addressing different governance levels; strategical (e.g. general and large scale visions and plans), tactical (e.g. a city district park programme), operational (e.g. yearly management plan of a green space) or reflexive (e.g. monitoring routines for a certain species) (Frantzeskaki and Tilie 2014; Kabisch 2015).

Property rights

One institutional component that is highly relevant in urban settings is represented by property rights – mechanisms used to control the use of a property (e.g. natural resources such as land, fish stocks, water supply) as well as the behaviour among people in relation to this property (Bromley 1991). Ostrom and Schlager (1996) identified five different property rights: right of access, right of extraction, right of management, right of exclusion and right of alienation. A person can have one or several of these rights. Property rights can be seen as problem solving devices that can support conflict resolution processes, provide transparency and build trust among property users (Bromley 1991). On the other hand, they can also become barriers to change, flexibility and inclusion, and thereby hinder adaptation to new circumstances, also known as path dependency (Pierson 2004). Property rights are closely related to ownership, where the property rights are dependent on type of ownership as well as the rights and duties of the owner.

Urban land use, governance and biodiversity

The characteristics of modern cities, such as land use patterns, are results of their governance – the interactions between actors and institutions at multiple levels and over time (Agrawal and Ostrom 2006). The land use patterns are expressions of these interactions that in turn are shaped by historic legacies as well as related political priorities and applied steering mechanisms. Politics, polity and policies of this governance are also the means to steer the urban development towards certain priorities.

Land is a key limiting factor in cities – the resource at stake – and urban land use is often governed by a set of strong formal institutions, such as urban planning regulations and property rights. An important component of urban planning is to secure common goods that need to be provided to the residents, such as infrastructure for transport, energy and water. Biodiversity is another common good to be safeguarded in the planning processes in cities and other landscapes (c.f. Zachrisson 2009). Since maintenance of biodiversity requires functional ecological processes, there is a dependence on land for these processes to take place, be it on the ground, or on vertical or multilayer surfaces as in modern greening of built structures. Therefore, urban biodiversity conservation activates the land use institutions at play in cities. However, essential to urban biodiversity conservation is not only the combinations of such formal regulations, but also praxis and informal norms that, through planning, design, management and use, impact urban green structures.

To work towards urban biodiversity conservation, it is important to understand the historical legacies and actor–institution interactions in order to identify needs and mechanisms for change. Such a *governance analysis* includes answering numerous questions (Figure 10.1) such as: what are the key actors involved in this green space – the stakeholders (e.g. land owners, users, decision-makers)?; what regulations steer the green space (e.g. laws regulating urban planning, nature conservation, forestry, agriculture etc.)?; what policies are in place (e.g. management plan, biodiversity strategy, green infrastructure program, detailed exploitation plan, clean water policy) and what are the main uses (e.g. bird watching, walking, playing, gardening)?; and how all this changes over time? (Frantzeskaki and Tilie 2014). Similar to urban green spaces being spatially embedded in the wider urban landscape, the governance of a green space is also part of a multilevel governance of a city district or a whole city that impacts the sustainability of the area and its capacity for biodiversity conservation. Therefore, also the wider governance context, beyond a specific site, must be included in the analysis (Bulkeley and Betsill 2005).

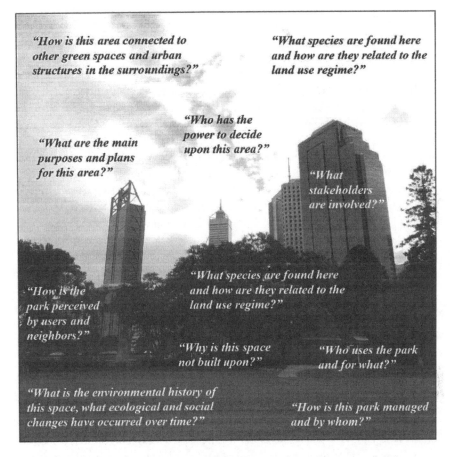

"How is this area connected to other green spaces and urban structures in the surroundings?"

"What species are found here and how are they related to the land use regime?"

"Who has the power to decide upon this area?"

"What are the main purposes and plans for this area?"

"What stakeholders are involved?"

"How is the park perceived by users and neighbors?"

"What species are found here and how are they related to the land use regime?"

"Why is this space not built upon?"

"Who uses the park and for what?"

"What is the environmental history of this space, what ecological and social changes have occurred over time?"

"How is this park managed and by whom?"

FIGURE 10.1   An urban green space in Perth, Australia through a governance lens with questions that can be answered through a *governance analysis*. (Photo courtesy: Alessandro Ossola.)

Protection of nature in cities – a biocentric approach

"Should we protect this environment for a threatened species or build houses for those who do not have a home in our city instead"? This question illustrates a common dichotomy between biocentric and anthropocentric arguments in urban planning. To set aside land is the main strategy for biodiversity conservation worldwide and a strategy that goes back more than 100 years in many countries (for example in the US and Sweden). It originates from the realisation of the negative environmental impacts of nineteenth century industrialisation. Nature had to be protected from such negative human impacts and the idea of nature conservation by protecting areas was born (Adams 2006). The legacy of protecting nature from humans is still evident in policy implementation of nature conservation in urban landscapes, where an either-or thinking is prevalent, as if there were no interim alternatives (Erixon *et al.* 2013).

A strong formal institution for biodiversity

Establishment of nature protected areas is usually based on formal institutions, such as national environmental protection legislation and/or international legislations or agreements such as *European Union Habitat Directive*. These formal institutions formulated as legislations specify relatively well what the establishment entails and the procedures to follow in planning and management. Nature protected area establishment means radical changes in the institutional frames for how the land is used – the property rights – which in turn impacts both social and ecological aspects. One property right that might change is ownership whereby many establishment processes require that public sector actors buy the land from present land owners (e.g. Swedish national parks). However, depending on the format of protection, private land ownership can also be part of the protected area (e.g. Swedish nature reserves).

The first establishments of protected areas in the early 1900s followed the rationales of hierarchical steering where the national level decision was a top-down process and often with the main purpose of limiting the ability of people to use the area (Adams 2006). This has shifted towards a governance approach, where citizens' participation in planning and management is seen as an essential asset for a sustainable development of the protected area (Zachrisson 2009). Modern nature protection is more about involving people and providing commons to be used than keeping them out, but not without difficulties as discussed by Brown (2003).

The main goal of establishing a protected area is that the institutional changes of land use will result in a conservation of the identified values, often biological diversity. This objective has been especially strong since the ratification of the Convention of Biological Diversity in 1992 (Borgström *et al.* 2013). Biodiversity conservation activates a number of challenging issues relating to present ecological knowledge – "*What biodiversity is located where?*", "*What biodiversity should/needs to be conserved?*" and "*What is the best way of doing it?*" The bulk of knowledge concerns the species level of biodiversity for which there is the longest monitoring records (compared to genetic and landscape biodiversity). Therefore, biodiversity conservation generally rests on the knowledge of species and their habitats. Threats to habitats and decreasing species populations are key factors in prioritising what areas to protect (Emneborg and Götmark 2000) and the main source of information is the IUCN red list and similar national lists of threatened species (www.iucnredlist. org). Setting aside land for biodiversity conservation is a very strong instrument for steering land use within a landscape, and, therefore, the arguments need to be very solid and convincing, such as an area being essential for the survival of a certain species (Emneborg and Götmark 2000, Borgström *et al.* 2013). This is especially pronounced in urban landscapes where land is a limited and expensive resource.

In addition to species diversity, many countries motivate nature protection with goals of conserving a representation of their national biotopes. Furthermore, the objectives of protected areas are changing over time and change differently depending on the local context (Borgström *et al.* 2013). Besides biodiversity conservation, other objectives used include recreational, economic, cultural and educational purposes (ibid.). At present in Sweden, it seems to be easier to find strong support for biodiversity conservation compared to other arguments for nature protection, such as recreation and human health benefits, where knowledge and juridical frameworks are less developed (Stenseke and Hansen 2014).

The urbanisation of nature protection

In the mid 1990s the scientific field of urban ecology entered a phase of strong development (McDonnell 2011; Pickett *et al.* 2011). It has been recognised that many cities are located in biodiversity rich areas (Cincotta *et al.* 2000) and therefore, following the agreements of the Convention of Biological Diversity, there is a need to find strategies for how to protect this biodiversity given the rapidly expanding cities around the world (Trzyna 2007, 2014). There was also a call for biodiversity conservation to focus on landscapes with a more intensive human use, e.g. agricultural lands and cities (Dunn *et al.* 2006; Kowarik 2011). Nature protection areas in cities are also increasingly part of a wider discussion of urban sustainable development (Trzyna 2014) and, consequently, their purposes are diversifying. Besides preserving certain species or biotopes, these areas are generally established to halt urban expansion, to serve as recreational grounds and to provide ecosystem services. Urban nature conservation, therefore, includes many different institutions and actors in a complex governance system.

Present nature protection areas in cities have different relations to urbanisation. In some cases, previously remote protected areas have been embedded into the expanding urban fabric. McDonald *et al.* (2008) estimated that one quarter of the world's protected areas are located within 17 km of a city and that this distance is likely to decrease in the future. In other cases, there are political ambitions leveraging biodiversity conservation through the establishment of protected areas. In 2003, the Swedish government decided formal protection of nature should be the main instrument for securing urban biodiversity, resulting in an increase in the rate of establishment of nature reserves in the Swedish cities. Such establishment of new protected areas to safeguard urban biodiversity within cities means that the arguments for biodiversity enter the urban planning arena – the "*what to do with the urban land*" market. This means that the formal and informal institutions of urban land use planning meet the institutions of biodiversity conservation. It is common that this meeting of institutions creates a dichotomy between development and conservation, where conservation is framed as being against development of new housings or infrastructure (Erixon *et al.* 2013). The assumption is that development and conservation cannot support each other and that the boundaries of the protected areas also are the limits to what can be developed. The protected area is seen as something different that has to be handled separately in the planning process. This is amplified by the separate knowledge spheres of professions involved in urban landscape governance, for example between urban planners and ecologists. This separation of urban nature protected areas from the rest of the urban landscape is also exemplified by the numerous reports on urban biodiversity and ecology that are inherently difficult to integrate into the general comprehensive or detailed urban planning (Sandström *et al.* 2006).

What kind of urban biodiversity to prioritise?

The recognition of the importance of urban biodiversity has led to a rather unconsidered application of existing nature conservation governance structures, institutions and procedures, including a legacy of nature protection concepts such as *wilderness* and *pristine nature*. Given that cities present different social–ecological conditions than other landscapes this application results in difficulties that might even lead to negative effects on biodiversity (ibid.). In the following section, such issues of conventional nature conservation in cities are discussed.

Urbanisation alters physical and ecological patterns and processes, for example, species presence, behaviour and composition, hydrology, climate and disturbance regimes (Niemelä *et al.* 2011). Urban species are to a large extent generalists that are adaptive and less sensitive than specialist species, which have very specific habitat requirements and therefore are less common in cities (Werner 2011). Existing specialist species might also be remnants from former land uses that cannot be supported by the present urbanising landscape in the long run (Foster *et al.* 2003). This urban situation questions the basic foundation for prioritisation in conventional biodiversity conservation, where a focus on specialised (and/or threatened, red-listed) species might result in rather few protected areas and that these areas in the end are insufficient to protect specialist species over time (Dearborn and Kark 2009). On the other hand, when present, red-listed/threatened species that are targeted by legislation provide strong arguments needed for setting aside land in cities. For example in Stockholm, Sweden, several protected areas have been established where the existence of a species targeted by the EU Habitat Directive has been decisive and changed major urban development plans.

Another consequence of the specifics of urban ecology is the spatial location of biodiversity in cities. It has been shown that harbours and waste deposits are especially rich in biodiversity due to the influx of new species (Kowarik 2011). However, not all of these new species are desirable because they might be invasive and thereby threaten native biodiversity. Cities are also places where nature is created and hence novel ecosystems are developed (for example water-sensitive urban designs) (ibid.). However, these new species and environments are seldom recognised by nature conservation due to the focus on preserving existing values under threat and preferably in already identified green spaces (Perring *et al.* 2013). As a consequence, large parts of the overall urban biodiversity are not captured or accounted for when applying conventional nature conservation frameworks in cities.

Another issue of conventional biodiversity conservation is that urban green spaces that hold biodiversity are usually small and isolated by surrounding dense built up areas. Therefore, in order to conserve urban biodiversity there is a need for safeguarding linkages between those areas – also known as the green infrastructure (Ahern 2013). However, these linking green structures might only encompass common species, and therefore do not gain the strong legal support needed to motivate a formal protection. There is a generalised lack of formal institutions that target these linkages grounded in landscape ecology principles, which is especially troublesome in large cities (ibid.). It is uncommon that the whole green infrastructure, including both green spaces of different types and sizes, and linkages between those spaces, is protected by formal institutions. On the contrary, this green infrastructure is often partly protected and partly not, which can cause confusion about the property rights. In many cases the green spaces in the proximity of protected areas are very attractive for urban developments due to their higher land values (Wittemyer *et al.* 2008; Borgström *et al.* 2011). The combination of informal institutions such as soft recommendations of green infrastructure preservation and formal institutions of protection of specific areas within these structures might even contribute to the isolation of green spaces, increasing difficulties for biodiversity conservation.

These concerns of the implementation of conventional biodiversity conservation in urban contexts presented here lead to questions about what kind of urban biodiversity needs to be preserved, by what arguments and means. It seems that the current practices are trapped in the legacy of biodiversity conservation principles developed for other types of landscapes. Hence, there is a need for revised or new institutions that provide strong arguments for landscape connectivity, novel ecosystems and biodiversity hotspots to establish protection for the larger urban green infrastructure. Another approach is to find convincing arguments and ideas that carry the values of biodiversity into the urban planning and management without the need of strong formal institutions such as protected areas and listed species. Some argue that this approach is the main added value of working with ecosystem service frameworks (Borgström 2013).

Negotiated locations and boundaries

In Sweden and in other countries, the process of establishing protected areas has moved from being only a national, top-down directive to become the responsibility of local decision makers. At the local scale, decision makers often have many different responsibilities and common interests to balance, where biodiversity is one. In a study of Swedish nature reserves it was shown that this has an impact on where and what kind of nature is protected, if any (Borgström 2009). Protected areas in cities tend to be located on land that is the least interesting for exploitation, where there is a large part of land owned by the public sector, and/or where there is a strong local opinion (ibid.; Ernstson and Sörlin 2009). Furthermore, the identification of potential areas is dependent on ecological knowledge that is usually based on a combination of expert inventories and local information (Yli-Pelkonen and Kohl 2005), and perceived threats to identified values that need protection (Emneborg and Götmark 2000). This means that areas are rarely selected based on a comprehensive assessment of biodiversity rich sites and their connectivity in the larger landscape, but rather from scattered information developed over time and in relation to emergent threats and public engagement.

The boundaries of a protected area are commonly decided by a compromise between several parameters such as ownership, land use history, species presence and surrounding stakeholders. In an urban land use map, the protected areas are commonly demarked by a solid line that represents an institutional boundary (Figure 10.2). This line divides the landscape based on where different property rights are located – the *where and who has the right to do what*. It is a boundary that also can be traced in the administration and management, where the protected area is handled for example by a nature conservation division (e.g. ecology experts) at the public authority and the land outside this area by a diversity of private and civic stakeholders and other divisions at the public authority (e.g. planning, road and park management). The line is also easily interpreted as an indicator of where in a particular landscape section biodiversity is a priority (within the boundary) and where it is less important/valuable (outside the boundary) (Borgström *et al.* 2011). In the extension the map with protected areas might give the impression that the land use within these boundaries will guarantee biodiversity. These interpretations show the backside of applying strong formal institutions in the contested urban landscape and call for a need to translate frameworks and tools of biodiversity conservation to better fit urban social–ecological dynamics.

FIGURE 10.2 Igelbäcken nature reserve marked by a solid line in the green wedge of Järva in northern Stockholm, Sweden (Source: Orthophoto modified from Google Maps).

Ecosystem services in urban planning – an anthropocentric approach

Ecosystem services is a concept developed within the field of natural resource management in the 1990s as a way to illustrate, communicate and investigate the linkages between ecological processes and social well-being (Braat and de Groot 2012). Since then, research about ecosystem services has flourished and has developed within a large diversity of fields besides natural resource management, e.g. ecological economics, conservation biology, urban planning and design and sustainability science (Costanza and Kubiszewski 2012). At the same time, it has become an important global policy instrument for working with strong sustainable development, where the capacity of the biosphere sets the frame for human activities (MEA 2005). Many countries have recently incorporated the concept of ecosystem services as part of their processes to achieve the goals of the Convention of Biodiversity (www.cbd.int/sp/targets). In relation to urban biodiversity, the concept of ecosystem services has been adopted both in biodiversity conservation, as yet another argument for protecting nature (Niemelä *et al.* 2010), and in urban planning as a way to handle environmental challenges (regulating ecosystem services) and add value to built-up areas (cultural ecosystem services) (Niemelä *et al.* 2010; Lorance Rall *et al.* 2015). The concept of ecosystem services is, similar to protected areas, a tool for working with urban biodiversity conservation. However, so far, the concept lacks a clear governance profile which is urgently needed in the ongoing implementation, not the least within urban planning (Loft *et al.* 2015, Kabisch 2015).

Planning for urban common goods

Urban planning is similar to nature protection in that it is a strong, formal institutional framework for handling land use decision-making processes. It has a long tradition in many countries and serves as the main mechanism for safeguarding the provision of basic, common services to the densely settled urban residents, such as housing, food, water and energy supply, and sanitation and waste management. In many cities around the world these goals are still to be met, which is a result of malfunctioning organisation of urban land use, decision making and service provision. Further, the provision of areas for recreation, such as green space, is often part of the urban planning ambitions. In many countries, urban planning follows strict protocols regulated by law that state i) what the resulting plans must include, ii) how the planning process should be designed and, iii) how and when certain actors should be involved. Urban planning is closely linked to democratic principles of decision making and the plans are approved by elected politicians. Similar to nature protection, urban planning has increasingly acknowledged the importance of citizen's participation for successful planning outcomes, and therefore urban planning can be described as based on both representative and deliberative processes.

Nature for humans in urban planning

As stated in the above section about urban nature protection, the value of bio-diversity has inherently had difficulties in being acknowledged among the many interests that are negotiated within urban land use planning. Therefore, the eco-system services approach has been appreciated for creating grounds for dialogues among planning, environmental management and nature conservation profession-als (Borgström 2013). The common denominator among these professionals is the goal to form a landscape for providing services supporting human welfare over time, and through the ecosystem services concept nature becomes a clear con-tributor to that goal. In addition, this anthropocentric approach to urban nature makes it easier to relate to for laypersons, compared to the biocentric approach. For example, the importance of green spaces for recreation and climate change adaptation is more directly beneficial to a person than protection of a rare species as part of biodiversity.

Another benefit of the ecosystem services concept is that it expands from the focus on green spaces where the ecosystem services are generated to also include areas where they are needed and used, such as the built environment (Andersson et al. 2015). This potentially leads to more integrated views of urban nature where processes, linkages and flows in the whole landscape are recognised (Andersson et al. 2014). Compared to conventional nature conservation, the ecosystem services approach includes landscape linkages as well as the new and restored nature, and therefore, potentially captures a larger proportion of the urban biodiversity.

However, in practical planning there are several traps that threaten these potentials. For instance, looking at the most progressive municipal planning processes with regards to ecosystem services in Sweden, there is still a lot of map-ping of green space and identification of what ecosystem services they provide, with very limited recognition of where those services are needed (Borgström 2013). Furthermore, the work with ecosystem services in urban planning is also to a large degree driven by data availability, which results in a subset of services included in the analysis (Primmer and Furman 2012). What then becomes cru-cial is an assessment investigating what green structures are needed where to provide ecosystem services (Burkhard et al. 2012). In other words, the supply side of the ecosystem services approach prevails over the demand side in present day urban planning.

Even if a city has invested in analysis of ecosystem services to inform their urban planning, the material is in many cases not well integrated into the actual plan-ning process and decision making (Delshammar 2015). This shows how strong the formal institutions of urban planning are, where it is difficult to integrate new aspects within existing frameworks. The concept of ecosystem services is well understood in urban planning (Figure 10.3), but when moving into the mecha-nisms of operationalising the concept it becomes clear that there is a severe lack of knowledge and tools, especially regarding the governance aspects (Primmer and

Furman 2012; Borgström 2013; Delshammar 2015; Lorance Rall *et al.* 2015). As a consequence, the ecosystem services approach is limited to places where there is a will and interest among the urban actors (e.g. in a new eco-profiled development area), and not where it is needed the most in a city (e.g. in a socially deprived, polluted and noisy neighbourhood).

Ecosystem services in a governance vacuum

In the urban landscape characterised by a high degree of heterogeneity of different land uses, the provision and use of ecosystem services approach becomes a matter of who is to contribute to, and who has the right to enjoy common goods. This is especially true since the need for ecosystem services does not necessarily co-occur within publicly owned green spaces (Gomez-Baggethun *et al.* 2013). While protected areas are framed by institutions with clear locations and boundaries, the provisioning of ecosystem services includes green structures in many different institutional settings, with different profiles of actor involvement, ownership, accessibility and responsibility (Andersson *et al.* 2014; Loft *et al.* 2015). There are today very few examples of instruments and incentives for local urban actors other than the public sector to invest in ecosystem services, although payments for ecosystem services is a growing field of study in other types of landscapes.

The governance vacuum is also resulting from the breadth of the ecosystem services concept, covering, for example, food production, decomposition, micro climate regulation and spiritual values, and therefore lacks a clear organisational home in most of the current urban governance structures (de Groot *et al.* 2010; Scarlett and Boyd 2015). In Sweden, the national level work with ecosystem services is spread among a number of national agencies responsible for nature conservation, urban development and agriculture to mention a few. This can be compared to climate change adaptation and water security that also are examples of complex environmental challenges stretching many different sectors and levels. The present governance structure including organisations, institutions and decision-making procedures are not well suited to handle these kinds of cross-sectorial issues. In the efforts towards implementation and operationalisation of the ecosystem services approach collaboration is essential, but the collaborative arenas often suffer from the participants having unclear mandates, roles, responsibilities and commitments (Loft *et al.* 2015).

Multifunctional green spaces in dense cities – a governance challenge

One of the dominating urbanisation strategies is the densification of existing built up areas motivated by energy and transport efficiencies and sparing of land for biodiversity, recreation and food production (Lin and Fuller 2013). This densification results in a decrease of open and green spaces within cities. At the same time this decreasing amount of green spaces has to meet diversifying demands from an increasing population. In this situation, the purposes of protected areas are relatively clear and already prioritised, in comparison with other green spaces where different interests (e.g. ecosystem services) compete in a more unclear institutional situation.

An emerging green space strategy within urban densification is to promote multifunctionality of the remaining green spaces, where the provision of ecosystem services is maximised (Lovell and Taylor 2013). This raises a number of governance issues, such as which ecosystem services are most important for the site and its surrounding, which ecosystem services can/cannot be combined and by whom is the ecosystem services prioritised and with what motives (Madureira and Andresen 2014). From today's focus on the physical design of these multifunctional green spaces there is a need for paying attention to the design of their governance.

Independently of being an urban protected area or a green space with another institutional framing, the increase of population density will require more careful planning and management to support long-term multifunctionality. Funding is already a limiting factor of urban green space management that will be even more pronounced in densifying cities, which call for development of new management arrangements for securing funding, monitoring, communication and education (Barthel *et al.* 2010; de Groot *et al.* 2010).

FIGURE 10.3 *Ecosystem services in urban planning – a guide.* The c/o city project (www.cocity.org) has developed several tools and models for working with ecosystem services. How these will be used and how they will contribute to urban biodiversity and sustainable development largely depends on the capacity of the governance structures to adapt and transform in relation to these new approaches. Picture: Front page with Pocket park by White 2014.

Concluding remarks

This chapter has, by the use of two approaches for working with urban biodiversity conservation, i.e., conventional nature conservation by formal protection and ecosystem services in urban planning, illustrated the multifold aspects of governance that urban biodiversity actors have to take into account. Both the biocentric and the anthropocentric perspective on biodiversity as a common good require that the political processes, institutional structures and policy content of urban landscape governance be addressed in order to successfully support urban biodiversity. The conventional nature protection approach is trapped in the rare and threatened species conundrum whereby a large proportion of the urban biodiversity is not captured. Furthermore, there is a lack of measures within this approach for targeting linkages between green spaces, which is essential for overall landscape connectivity and biodiversity. The intrinsic value of nature has been shown to have large difficulties as an argument in competition with other common goods within urban planning. However, when the protected areas have been established the institutional setting is clear. The anthropocentric approach represented by the ecosystem services concept has shown potential in integrating the urban biodiversity into urban planning, including a larger proportion of the urban biodiversity as well as the landscape linkages. However, the green structures providing these services are scattered in a very heterogeneous governance structure of multiple institutions creating many uncertainties. There are presently very few examples of mechanisms that require local urban actors to contribute to an urban common good such as ecosystem services provision. This likely causes a mismatch between where ecosystem services can be provided and where they are needed in a city. For addressing these governance challenges there is a need for well-designed knowledge co-creation processes where public, private and civic actors in collaboration with researchers jointly develop and test solutions adapted to the specific contexts and that seek to improve the urban governance to better support urban biodiversity.

Acknowledgements

This chapter has been written with the support from the CONATURE project funded by Formas, the ISSUE project funded by Formas and the C/O CITY project funded by VINNOVA. It has been developed with inspiration from discussions with Dr Erik Andersson, Dr Annika Dahlberg and partners within all three projects.

References

Adams W.M. (2006) *Future Nature: A Vision for Conservation*. Earthscan, London.
Agrawal A. and Ostrom E. (2006) Political science and conservation biology: A dialog of the deaf. *Conservation Biology* 20(3): 681–682.
Ahern J. (2013) Urban landscape sustainability and resilience: The promise and challenges of integrating ecology with urban planning and design. *Landscape Ecology* 28: 1203–1212.

Andersson E., Barthel S., Borgström S. Colding J., Elmqvist T., Folke C. and Gren Å. (2014) Reconnecting cities to the biosphere: Stewardship of green structure and urban ecosystem services. *Ambio* 43: 445–453.

Andersson E., McPhearson T., Kremer P., Gomez-Baggethun E., Haase D., Tuvendal M. and Wurster D. (2015). Scale and context dependence of ecosystem service providing units. *Ecosystem Services* 12: 157–164.

Barthel S., Folke C. and Colding J. (2010) Social-ecological memory in urban gardens: Retaining the capacity for management of ecosystem services. *Global Environmental Change* 20: 255–265.

Berkes F. and Folke C. (1998) *Linking Social and Ecological Systems: Management Practices and Social Mechanisms for Building Resilience.* Cambridge University Press, Cambridge.

Borgström S. (2009) Patterns and challenges of urban nature conservation: A study of southern Sweden. *Environment and Planning A* 41: 2671–2685.

Borgström S. (2013) Appendix 4: Ekosystemtjänstperspektivet i svensk miljöpolicy och praktik – potentialer, barriärer och vägar mot integration, pp. 245–258 in SOU 2013:68 Synliggöra värdet av ekosystemtjänster. Åtgärder för välfärd genom biologisk mångfald och ekosystemtjänster.

Borgström S., Cousins S. and Lindborg R. (2011) Outside the boundary: Land use changes in the surroundings of urban nature reserves. *Journal of Applied Geography* 32: 350–359.

Borgström S., Lindborg R. and Elmqvist T. (2013) Nature conservation for what? Analyses of urban and rural nature reserves in Southern Sweden 1909–2006. *Landscape and Urban Planning* 117: 66–80.

Braat L. and de Groot R. (2012) The ecosystem services agenda: Bridging the worlds of natural science and economics, conservation and development, and public and private policy. *Ecosystem Services* 1(1): 4–15.

Bromley D.W. (1991) *Environment and Economy: Property Rights and Public Policy.* Blackwell, Oxford.

Brown K. (2003) Three challenges for a real people-centred conservation. *Global Ecology and Biogeography* 12(2): 89–92.

Bulkeley H. and Betsill M. (2005) Rethinking sustainable cities: Multilevel governance and the 'urban' politics of climate change. *Environmental Politics* 14: 42–63.

Burkhard B, Kroll F., Nedkov S. and Muller, F. (2012) Mapping ecosystem service supply, demand and budgets. *Ecological Indicators* 21: 17–29.

Carlsson L. and Berkes F. (2005) Co-management: Concepts and methodological implications. *Journal of Environmental Management* 75: 65–76.

Cincotta R. P., Wisnewksi J. and Engelman R. (2000) Human population in the biodiversity hotspots. *Nature* 404: 990–992.

Costanza, R. and I Kubiszewski. 2012. The authorship structure of "ecosystem services" as a transdisciplinary field of scholarship. *Ecosystem Services* 1(1): 16–25.

Dearborn D. and Kark S. (2009) Motivations for conserving urban biodiversity. *Conservation Biology* 24(2): 432–440.

Delshammar T. (2015) Ecosystem services in municipal spatial planning. Progress report, Sweden 2015. Report 2015:23. SLU, Alnarp.

de Groot R.S., Alkemade R., Braat L., Hein L. and Willemen L. (2010) Challenges in integrating the concept of ecosystem services and values in landscape planning, management and decision making. *Ecological Complexity* 7: 260–272.

Dunn R.R., Gavin C.G., Sanchez M.C. and Solomons J.N. (2006) The pigeons paradox: Dependence of global conservation on urban nature. *Conservation Biology* 20: 1814–1816.

Emneborg H. and Götmark F. (2000) The role of threat to areas and initiative from actors for establishment of nature reserves in southern Sweden 1926–1996. *Biodiversity and Conservation* 9: 727–738.

Erixon H., Borgström S. and Andersson E. (2013) Challenging dichotomies: Exploring resilience as an integrative and operative conceptual framework for large-scale urban green structures in Stockholm, Sweden. *Planning Theory and Practice* 14(3): 349–372.

Ernstson H. and Sörlin S. (2009) Weaving protective stories: Connective practices to articulate holistic values in the Stockholm National Urban Park. *Environment and Planning A* 41: 1460–1479.

Folke C. (2006) Resilience: The emergence of a perspective for social–ecological systems analyses. *Global Environmental Change* 16: 253–267.

Foster D., Swanson F., Aber J., Burke I., Brokaw N., Tilman D. and Knapp A. (2003) The importance of land-use legacies to ecology and conservation. *BioScience* 53(1): 77–88.

Frantzeskaki N. and Tilie N. (2014) The dynamics of urban ecosystem governance in Rotterdam, the Netherlands. *Ambio* 43: 542–555.

Fung A. and Wright E.O. (2001) Deepening democracy: Innovations in empowered participatory governance. *Politics Society* 29(1): 5–41.

Gomez-Baggethun E., Kelemen E., Martin-Lopez M., Palomo I. and Montes C. (2013) Scale misfit in ecosystem service governance as a source of environmental conflict. *Society and Natural Resources* 26: 1202–1216.

Grimm N.B., Faeth S.H., Golubiewski N.E., Redman C.L., Wu J., Bai X. and Briggs J.M. (2008) Global change and the ecology of cities. *Science* 319: 756–760.

Imperial M. (1999) Institutional analysis and ecosystem-based management: The institutional analysis and development framework. *Environmental Management* 24(4): 449–465.

Kabisch N. (2015) Ecosystem services implementation and governance challenges in urban green space planning: The case of Berlin, Germany. *Land Use Policy* 42: 557–567.

Kooiman J. (2003) *Governing as Governance*. SAGE, London.

Kowarik I. (2011) Novel urban ecosystems, biodiversity, and conservation. *Environmental Pollution* 159: 1974–1983.

Lange P., Driessen P.P.J., Sauer A., Bornemann B. and Burger P. (2013) Governing towards sustainability: Conceptualizing modes of governance. *Journal of Environmental Policy & Planning* 15(3): 403–425.

Lin B. and Fuller R. (2013) Sharing or sparing? How should we grow the world's cities? *Journal of Applied Ecology* 50: 1161–1168.

Loft L., Mann C. and Hansjurgens B. (2015) Challenges in ecosystem services governance: Multi-levels, multi-actors, multi-rationalities. *Ecosystem Services* 16: 150–157.

Lorance Rall E., Kabisch N. and Hansen R. (2015) A comparative exploration of uptake and potential application of ecosystem services in urban planning. *Ecosystem Services* 16: 230–242.

Lovell S. and Taylor J.R. (2013) Supplying urban ecosystem services through multifunctional green infrastructure in the United States. *Landscape Ecology* 28: 1447–1463.

Madureira H. and Andresen T. (2014) Planning for multifunctional urban green infrastructures: Promises and challenges. *Urban Design International* 19: 38–49.

McDonald R.J., Kareiva P. and Forman R.T.T. (2008) The implications of current and future urbanization for global protected areas and biodiversity conservation. *Biological Conservation* 141: 1695–1703.

McDonnell M.J. (2011) The history of urban ecology: An ecologist's perspective. In Niemelä J. (ed.), *Urban Ecology: Patterns, Processes, and Applications* (pp. 5–13). Oxford University Press, New York.

MEA (2005) *Ecosystems and Human Well-being: Current State and Trends*. Island Press, Washington, DC.

Niemelä J., Saarela S-R., Söderman T., Kopperoinen L., Yli-Pelkonen V., Väre S. and Kotze J. (2010) Using the ecosystem services approach for better planning and conservation of urban green spaces: A Finland case study. *Biodiversity and Conservation* 19: 3225–3243.

North D. (1991) *Institutions, Institutional Change and Economic Performance*. Cambridge University Press. Cambridge.

Ostrom E. and Schlager E. (1996) The formation of property rights. In Hanna S., Folke C. and Mäler K-G. (eds), *Rights to Nature* (pp. 127–157). Island Press. Washington, DC.

Perring M., Manning P., Hobbs R., Lugo A., Ramalho C. and Standish R. (2013) Novel urban ecosystems and ecosystem services. In Hobbs R., Higgs E. and Hall C. (eds), *Novel Ecosystems: Intervening in the New Ecological World Order* (pp. 310–325). Wiley, Oxford.

Peters G.B. and Pierre J. (2004) Multi-level governance and democracy: A Faustian bargain? In Bache I. and Flinders M. (eds), *Multi-Level Governance* (pp. 75–92). Oxford University Press, Oxford.

Pickett S.T.A., Cadenasso M., Grove M., Boone C., Groffman P., Irwin E., Kaushal S., Marshall V., McGrath B., Nilon C., Pouyat R., Szlavecz K., Troy A. and Warren P. (2011) Urban ecological systems: Scientific foundations and a decade of progress. *Journal of Environmental Management* 92: 331–362.

Pierson P. (2004) *Politics in Time: History, Institutions, and Social Analysis*. Princeton University Press, Princeton, NJ.

Primmer E. and Furman E. (2012) Operationalising ecosystem services approaches for governance: Do measuring, mapping and valuing integrate sector-specific knowledge systems? *Ecosystem Services* 1: 85–92.

Sandström U., Angelstam P. and Khakee A. (2006) Urban comprehensive planning: Identifying barriers for the maintenance of functional habitat networks. *Landscape and Urban Planning* 75: 43–57.

Scarlett L. and Boyd J. (2015) Ecosystem services and resource management: Institutional issues, challenges and opportunities in the public sector. *Ecological Economics* 115: 3–10.

Stenseke M. and Hansen A. 2014. From rhetoric to knowledge based actions: Challenges for outdoor recreation management in Sweden. *Journal of Outdoor Recreation and Tourism* 7–8: 26–34.

Trzyna T. (2007) *Global Urbanization and Protected Areas*. InterEnvironment California Institute of Public Affairs, Sacramento, California.

Trzyna T. (2014) Urban protected areas. Profiles and best practice guidelines. IUCN Best practice protected area guidelines series no. 22. IUCN, Gland.

Werner P. (2011) The ecology of urban areas and their functions for species diversity. *Landscape and Ecological Engineering* 7(2): 231–240.

White (2014) Ekosystemtjänster i stadsplanering – en vägledning. c/o city project.

Wittemyer G., Elsen P., Bean W., Coleman A., Burton O. and Brashares J. (2008) Accelerated human population growth at protected areas edges. *Science* 321: 123–126.

Yli-Pelkonen V. and Kohl J. (2005) The role of local ecological knowledge in sustainable urban planning: Perspectives from Finland. *Sustainability: Science, Practice and Policy* 1: 3–14.

Zachrisson A. (2009) Commons protected for or from the people: Co-management in the Swedish mountain region? Doctoral Thesis. Dept of Political Science. Umeå University. Sweden.

11

MANAGING URBAN GREEN SPACES FOR BIODIVERSITY CONSERVATION

An African perspective

Sarel S. Cilliers, Stefan J. Siebert, Marié J. Du Toit and Elandrie Davoren

Introduction

In this chapter we use a broad definition of urban biodiversity that is useful in a multidisciplinary approach in biodiversity conservation as discussed by Farinha-Marques *et al.* (2011). It includes the variety of animal, plant and microbe species (including their genetic and functional diversity) and the interactions between them, as well as the diversity of habitats along a rural to urban gradient (variety of ecosystems). It is also important to acknowledge the social and cultural context of urban biodiversity as people introduce and use species for 'horticulture, forestry and landscaping' according to Farinha-Marques *et al.* (2011).

McDonnell and Hahs (2014) describe two ideologies at play in managing urban biodiversity for *creating biodiversity-friendly cities*, namely the conservation of an area's local native biodiversity and managing biodiversity for the benefit of people, i.e. ecosystem services. McDonnell and Hahs (2014) eloquently show that these two ideologies need to be balanced to achieve long-term success. Management is required to achieve win–win situations that neither overemphasize conservation by creating areas wherein people are largely excluded (nature wins), nor de-emphasize conservation by managing solely for ecosystem service delivery regardless of its effect on other fauna and flora (people win) (Figure 11.1). Overemphasizing urban nature conservation could lead to *bounded thinking* in which ecologically significant green areas are isolated from the rest of the city and where the urban residents are characterized as disturbances or threats to nature that need to be educated at all cost, instead of acknowledging that they may have alternative views of urban nature conservation that could be valued as *active citizen engagement* in managing urban biodiversity (Gill *et al.* 2009).

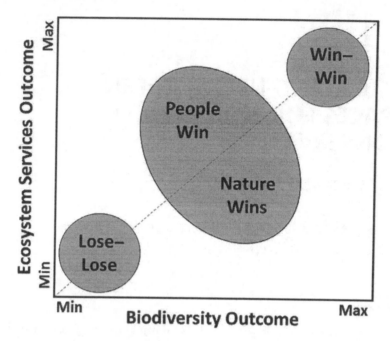

Moreover, in considering rationales for biodiversity conservation, Pearson (2016) identifies three different types of values, namely utilitarian, intrinsic and non-use values. He argues that the spatial extent and level of biological organization influences our perception and management of nature, and that the suitability and relevance of the different rationales for conservation change along these scales. For example, the existence value of a species such as any endangered species (e.g. tiger) has an intrinsic value on a global level, whereas populations of species, such as bees which serve as pollinators, have a significant local utilitarian conservation value based on the valuable ecosystem service they provide. Conservation decisions on this local level considering utilitarian values are relevant alternative arguments for biodiversity conservation, as local populations and genetic diversity will also be included (Pearson 2016). Hahs and McDonnell (2014) agree with this view and state that planning and management of cities, which focus on both humans and nature, need decisions on different spatial scales. This is particularly true when considering the specific urban ecological complexity that varies from species selection to strategies where the flow of nutrients, water, energy and organisms to and from the larger regions around cities is addressed.

According to a global study by Aronson *et al.* (2014), urban biodiversity (expressed as density of species) is best explained by anthropogenic features (land cover, city age) rather than by non-anthropogenic factors (geography, climate, topography). This seems to be even more pronounced in African cities (Balmford *et al.* 2001). In fact, the rapid expansion and prevailing management practices of African cities have a displacement effect on many green spaces and species (Fanan *et al.* 2011), which impedes the ability of urban green spaces to contribute to biodiversity conservation. To further exacerbate the plight of urban green areas in Africa, Muderere (2011) identified management-related factors that could be responsible, such as hostile urban land use policies (e.g. open spaces being converted to low cost housing), lack of enforcement of development control, political interference, resource shortages and outdated legislation.

Urban ecological studies in Africa show a clear bias towards South African cities. Applied urban ecological research in South Africa has followed several approaches (Cilliers 2010) focusing on different parts of the urban matrix, namely (1) fragmented natural areas in biodiversity hotspots, (2) public and private green spaces, and recently attempts to focus on the (3) whole urban green infrastructure (Cilliers and Siebert 2012). These three different approaches have led to different management practices to conserve urban biodiversity and are challenging the objective to find a balance between conservation of native biodiversity or biodiversity for human well-being, in different ways. In this chapter the three different approaches will be discussed.

Management of fragmented natural areas in biodiversity hotspots

Two of the largest cities in South Africa, Cape Town and Durban, are situated in biodiversity hotspots of international importance (Rebelo *et al.* 2011; Boon *et al.* 2016). Due to high human population growth rates resulting in large scale urban development in these cities, all native plant species are under extreme threat mainly due to transformation and fragmentation of their habitats (Rebelo *et al.* 2011; Boon *et al.* 2016). Biodiversity conservation in these cities focuses therefore on the natural remnants inside the cities as well as natural areas on the city margin. An approach of systematic conservation planning (SCP) is often followed that consists of two phases, namely systematic conservation assessment identifying spatially explicit priority areas for conservation including quantitative target areas, and the development of an implementation strategy and action plan (Driver *et al.* 2003).

The first phase of SCP, namely systematic conservation assessment, includes several data layers focusing on biodiversity, such as vegetation types, plant species distributions, rare and endangered species, ecological and evolutionary processes, habitat transformation, pressure of future land-uses and current protected areas (Driver *et al.* 2003; Ground *et al.* 2016). The SCP output for Cape Town is a biodiversity network (BioNet) that indicates current protected areas as well as critical biodiversity areas (CBAs) and critical ecological support areas (Holmes *et al.* 2012). A similar biodiversity plan was developed for Durban (Ground *et al.* 2016). The second phase of SCP is to implement the biodiversity networks (plans), by following a combination of different strategies (Holmes *et al.* 2012; Boon *et al.* 2016). One strategy is to implement spatial planning tools where high biodiversity areas are regarded as just as important as the other land uses (Boon *et al.* 2016). Another strategy is to use biodiversity impact assessments to ensure that CBAs in good condition are not developed, and a last strategy is the acquisition of land to be included in current formally protected areas (Holmes *et al.* 2012; Boon *et al.* 2016). Urban and peri-urban land acquisition is generally expensive, and therefore an additional implementation strategy of developing biodiversity stewardship partnerships between local government and landowners has been followed, empowering the owners to manage their land in exchange for reduced rates, tax relaxation and management advice (Holmes *et al.* 2012; Boon *et al.* 2016). A major challenge in the implementation of SCPs in South African urban areas is an issue of scale in that most existing plans are too coarse to be functional while fine-scale patch-related biodiversity issues are often ignored as they were not studied or the information is not available in a spatially explicit format (Ground *et al.* 2016).

Managing the fragmented natural areas in SCPs in South African cities is quite similar to managing natural areas away from cities and includes methods to simulate natural disturbances at historic levels while anthropogenic disturbances should be excluded according to Ground *et al.* (2016). The most common management practices followed are block burning and the removal of alien invader species (Van Wilgen *et al.* 2012; Ground *et al.* 2016). The management of fire-adapted natural areas in cities is extremely challenging due to conflicting issues causing tension between people with contrasting world views. This is the case of management practices such as prescribed burning that sustains the ecological integrity of the system as compared to the prevention and suppression of wildfires to protect humans and their properties. The fire management plan of the Table Mountain National Park on the outskirts of Cape Town had to include the city and surrounding areas which has led to a shift in focus from conservation to safety (Van Wilgen *et al.* 2012).

Management strategies to protect and restore native vegetation through the removal of alien invasive species often cause tension between conservationists and citizens. This is particularly evident if these species are highly valued for their cultural ecosystem services such as the pine forests in Cape Town (Van Wilgen *et al.* 2012). To address this tension, an invasive species management framework was proposed, in which some invasive species can be tolerated depending on their perceived benefits by urban residents which are measured against their perceived negative impacts on the environment (Gaertner *et al.* 2016). This framework includes a decision support system in which different stakeholders are involved to ensure that the choices which are made in terms of tolerance or eradication of the invasive species are *logic and defensible* (Gaertner *et al.* 2016).

In the Cape Peninsula region much tension is further caused by human–baboon conflicts and management actions should be sensitive to both human needs and humane treatment of baboons (Hoffman and O'Riain 2012). The proliferation of household pets can also increase tensions and threats to biodiversity as seen with the invasion of domestic cats into fragmented natural areas (Tennent *et al.* 2009). On the one hand, they control vermin such as rats, but they also prey on other species; a management issue then further complicated by the myriad opinions of citizens towards domestic cats, as an example from the KwaZulu-Natal province has shown (Tennent *et al.* 2009).

Overharvesting of natural resources from fragmented natural areas within the urban matrix can further contribute to the decline of biodiversity. Westernization and the mixing of cultures have resulted in the breakdown of cultural taboos that have historically protected biodiversity within cities. In Maputo, Mozambique, the sacred groves of the Licuati Forest are threatened by over logging of keystone species for charcoal production due to disrespect of ethnic custodianship (Izidine *et al.* 2008). Urban demand for medicinal plants has further exacerbated the loss of species. Although conservation strategies are widely promoted, there is increasing evidence of unsustainable harvesting practices in urbanized centres, such as ring barking of medicinal trees by commercial harvesters (Matsiliza and Barker 2001). All the examples above imply a reduction in overall species diversity and increased homogeneity relative to more properly managed and less disturbed sites.

Although ecosystem services are not the focus of SCPs, several studies have indicated that there is a correlation between biodiversity and ecosystem services (e.g. Egoh *et al.* 2009). Thus, attempts were made in Durban and Cape Town to link ecosystem services with the SCP plans. Davids *et al.* (2016) aimed to investigate potential management synergies between ecosystem services and biodiversity in Durban, South Africa, by identifying ecosystem services hotspots and determining whether they overlap with CBAs, conservation areas, and land ownership. The ecosystem services they investigated are related to ecosystem functions such as carbon storage, flood attenuation, run-off reduction, erosion prevention, and sediment trapping and retention of nutrients in runoff. Although this study identifies win–win opportunities (see Figure 11.1) for the co-management of biodiversity and ecosystem services, Davids *et al.* (2016) have also alluded to the challenges provided by the fact that most of the conservation areas are not multifunctional and that many of the ecosystem service hotspot areas are outside CBAs and conservation areas and inside private and communal lands. They are therefore more vulnerable to natural and anthropogenic disturbances. Various stakeholders need to be involved in a process of co-management of ecosystem services hotspots inside and outside of formally protected areas, including a prioritization process (Davids *et al.* 2016). O'Farrell *et al.* (2012) developed a rapid ecosystem service assessment for natural areas in Cape Town using the concept of ecosystem service bundles indicating which areas provide multiple services, but also mentioned that in certain areas biodiversity conservation should be prioritized above ecosystem services because of national responsibility towards and global concerns about the fragmented natural areas with high biodiversity in the hotspot areas.

Management of public and private green spaces

Public green spaces such as parks are highly managed, in contrast to fragmented natural areas found in cities of the developing world, and are considered as regulated initiatives, together with roadside greening (street trees), to increase green space (Swanwick *et al.* 2003). Their value largely lies in aesthetics and they are a common feature of well established, usually more affluent parts of urban centres. From a biodiversity point of view, urban parks contribute towards the biotic homogenization of cities, as the most tolerant and easily maintained lawn and tree species are generally globally recognized and promoted (Gong *et al.* 2013). This does not bode well for the conservation of native biodiversity in the urban environment. However, conservation initiatives increasingly promote the cultivation of native species and usually target public open spaces for such plantings (Anderson *et al.* 2014). For example, studies by Whitmore *et al.* (2002) of traffic islands in Durban, South Africa, have shown that the inclusion of indigenous shrubs, herbs and trees (i.e. high structural diversity due to different strata) increased invertebrate diversity in comparison with extensively mown areas (i.e. limited structural diversity). A survey conducted in Cape Town suggests that alien tree plantations have a reduced diversity of invertebrates when compared to urban gardens planted with indigenous plants (Pryke and Samways 2009). Both these cases of greening intervention highlight the importance of native plant species in the maintenance and enhancement of invertebrate diversity in urban green space (Anderson *et al.* 2014). The same applies for bird diversity, as the planting of native plant species with longtubed flowers could restore the nectarivorous bird guild in a city (Pauw and Louw 2012).

As public green spaces are managed by city councils and are generally funded by the taxpayer, co-management (or homeowner-based management) provides the opportunity for the community to become involved in the management of green space and can influence management decisions to benefit the user (Nkambule *et al.* 2016). This is important, as race, income and asset ownership are significant determinants of what residents consider as valuable green areas (Reuther and Dewar 2005). Successful co-management is dependent on environmental education, government involvement and events involving cross-cultural exchange (Graham and Ernstson 2012), such as the exemplary initiatives of the Sanitation, Beautification and Park Development Agency of Addis Ababa, Ethiopia (Mpofu 2013).

An important factor to consider in the management of public green spaces is the fear of crime. Studies indicate that people avoid many public green spaces due to fear of crime (Lemanski 2004). Another factor influencing public green spaces is public perception. Even those residents in favour of green open spaces have a wide variety of opinions on how these green spaces should look. McDonnell (2007) argues that there is a large difference between the way people perceive nature and ecological function. Tidy lawns, parks and gardens – ecologically 'sanitized' areas – are generally preferred, whereas more ecologically functional 'messy' landscapes with elements such as decaying leaf litter, dead branches, un-mown sidewalks and parks typically receive public criticism despite the important contribution to ecological processes (Ossola *et al.* 2016). Criticism of 'messy' landscapes is not only due to aesthetics but also because of fear for dangerous wildlife (e.g. snakes, spiders) and potential criminals. However, to enhance urban biodiversity 'messy' parks and gardens should be tolerated especially when this relates to 'the loss of niches caused by the ecological sanitization of the ground layer' (McDonnell 2007, p. 84). Tensions also arise when the public does not approve of the removal of invasive trees or ornamental plants from public green space and often even advance the spread of these species (Gaertner et al. 2016).

Biodiversity management should also contend with the detrimental effects of certain species, described as ecosystem disservices (Lyytimäki and Sipilä 2009). Examples of disservices include allergenic reactions (e.g. wind dispersed pollen) and falling trees and branches causing structural damage to properties (see Von Döhren and Haase 2015 for a list of disservices). Little research has been done on the perceived negative impacts of urban biodiversity in Africa (Von Döhren and Haase 2015), but this needs to be considered, especially in the management of public green spaces which are often in the most densely populated areas in cities. Species causing disservices need careful management, and decisions regarding their potential removal should be weighed against the beneficial services provided, as it was suggested in the previous section for natural areas in and around cities.

Private green areas such as domestic gardens have different management practices, financing mechanisms and decision makers compared to public green spaces (Shackleton 2006). Across Africa, gardening is very much needs–driven and depends on the ecosystem services that a community requires from their private spaces (Molebatsi et al. 2010; Idohou et al. 2014). Subsequently, managing private green spaces can determine both positive and negative impacts on biodiversity conservation depending upon the type and intensity of environmental management performed. It has been shown for Bujumbura in Burundi that private gardens play an important role in the functioning of the urban ecosystem and provide refuge to native plant and animal species (Bigirimana et al. 2012). However, gardens are also sources of invasive alien species that often outcompete native species (Alston and Richardson 2006).

In a South African study of private gardens along a socio-economic gradient, floristic diversity of such gardens increased with the availability of resources and intensity of garden management practices (Lubbe et al. 2010; Davoren et al. 2016). According to Loram et al. (2011), the variety of land covers and features (e.g. lawns, shrubs, flower beds, vegetable patches, ponds, patios, paths, sheds, greenhouses) will increase when people are interested in maintaining their own gardens. However, findings from South African cities suggest that the species diversity of gardens does not solely depend on garden management practices, but is also influenced by culture and socio–economic status (Lubbe et al. 2010; Davoren et al. 2016).

Studying private garden management practices might be the key to understand why people maintain biodiversity in cities, acknowledging the fact that gardens provide opportunities of practice to appreciate and utilize urban green areas. A study of domestic gardens in the North-West Province by Davoren (2016) has indicated management trends that hold direct relevance to the enhancement of biodiversity in South African cities and explains some of the potential mechanisms that drive the relationship between socio-economic status, behaviour of residents and actual impacts on habitats and biodiversity. However, it is important to acknowledge that people maintain biodiversity mostly for the sake of the ecosystem services that are provided (Cilliers *et al.* 2013) and rarely for the conservation value of rare species as such (Goddard *et al.* 2010). In this sense, the importance of gardens is predominantly one of habitat provision for the overall species pool (Sperling and Lortie 2010); and where rare and threatened species are maintained it should be seen as an additional benefit (Idohou *et al.* 2014).

Surveys in North-West Province revealed that, in poorer neighbourhoods, people cultivate plant species that are obtained from the surrounding natural areas or from friends and family. In the more affluent metropolitan areas, people planted what was available from commercial nurseries, what they obtained from friends and family, and in some cases selected plants based on gardening advice provided in gardening magazines and television shows. This social heterogeneity has a major implication for management practices as the type of plants cultivated will determine the watering and fertilization regime, pest and weed control. Hence, in the more affluent suburbs the *luxury effect* observed by Hope *et al.* (2003) prevails due to intensive garden management made possible by higher incomes. In affluent suburbs in metropolitan and urban areas, gardens are tended by part-time gardeners or gardening services (Table 11.1), which enable extensive plantings that enhances biodiversity through the creation of suitable habitat for species (Figures 11.2, 11.3a). In contrast, Davoren *et al.* (2016) found that the residents of lower income groups in deep rural, rural and peri-urban areas lacked the financial means to invest in garden management and channelled their limited resources to subsistence cultivation, which leads to a reduction in biodiversity (Figures 11.2, 11.3b).

TABLE 11.1 Outsourcing of garden activities (%) versus self-gardening in five settlements of South Africa along a rural–urban gradient.

	Tlhakgameng / Deep rural	Ganyesa / Rural	Ikageng / Peri-urban	Potchefstroom / Urban	Roodepoort / Metropolitan
Self-gardening	100	93	57	6	12
Gardening services	0	0	35	25	14
Hired gardeners	0	7	8	69	74

Self-gardening: owner of the garden is responsible for the upkeep;
Gardening service: owner hires a company to mow lawn and keep garden tidy;
Hired gardeners: owner hires a part-time gardener to conduct plantings, upkeep and garden development.

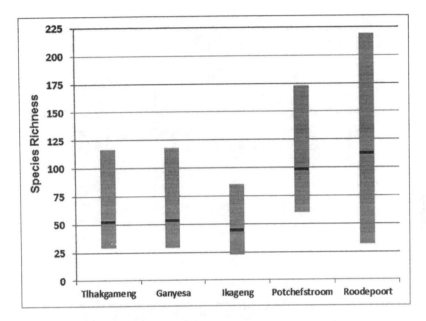

FIGURE 11.2 Average number of plant species recorded for gardens in five settlements along an rural–urban gradient in South Africa (indicated by solid black lines). Solid grey bars indicate the range in species richness recorded for gardens in each settlement.

Furthermore, the majority of the participants in the urban and metropolitan areas fertilize their gardens at least once a year, with chemical fertilizers, while the few residents that do fertilise their gardens in poorer areas use organic fertilizer (manure), which is easily obtained and cheaper than chemical fertilizers (Davoren 2016). Garden management practices do not only vary between different socio-economic status groups, but between different cultural groups as well. In Africa much of the management of green space is determined by indigenous knowledge systems (Idouhou *et al.* 2014). For instance, residents of the Tswana ethnic group believe that the immediate area around their houses should be open and devoid of any vegetation (Cilliers *et al.* 2009). They consider this as an indication of the tidiness of the household (*lebala concept*) (Figure 11.3c) and, subsequently, these areas are extensively swept and weeded (Molebatsi *et al.* 2010). This obviously leads to a reduction in biodiversity. In rural and deep rural areas residents maintain the *naga* (Figure 11.3d), a piece of natural vegetation within their yards, for medicinal or grazing purposes (Molebatsi *et al.* 2010), which in turn may enhance biodiversity within settlements considerably. This is also related to the biocultural diversity concept as suggested by Cocks (2006), as the cultural practices of inhabitants determine the usage and management of species in green space (Buizer *et al.* 2016).

(a)

(b)

(c)

(d)

FIGURE 11.3 Biodiversity and vegetation cover of gardens subjected to different management regimes based on cultural and socio-economic factors; a) westernized garden of wealthier suburbs; b) subsistence plots within gardens of peri-urban areas; c) *lebala concept* and d) *naga* surrounding an enclosure where livestock can be kept temporarily in rural and deep rural areas.

Management of the whole urban green infrastructure

From the literature it is clear that environmental management for urban biodiversity should not focus only on the fragmented natural areas or the public and private open spaces, but needs to include the entire urban matrix (Nilon 2011). According to Nilon (2011, p. 47), this approach can also be better understood by urban residents because the urban matrix contains 'the places where they work, live and play'. Conservation of biodiversity and its supporting ecosystem services is also an important strategy to plan resilient and sustainable cities (Ahern 2011). Sustainability and resilience are often used interchangeably, but in essence they are two complementing concepts (Elmqvist 2014; Meerow and Newell 2016). Sustainability is regarded as a normative concept and refers to the use and management of resources for current and future generations and is a critical goal for future urban development (Elmqvist 2014). The concept of resilience has been defined in numerous ways but is regarded as non-normative (Elmqvist 2014) and can in general be described as the ability of a system to react to change or disturbance without being inherently changed (Walker and Salt 2006). Resilience is a characteristic of a system in which humans and nature are coupled and which is dynamic, unpredictable and self-organizing. Thus, cities should be studied, planned and managed as complex social-ecological systems but on multiple scales and within a global context focusing on '*systems of cities to create a framework that manages resource chains for sustainability through resilience*' (Elmqvist 2014, p. 30, emphasis in original).

In terms of the contribution of biodiversity to build resilience in an urban system, the concepts of response diversity and redundancy need to be explained.

Response diversity refers to the variety of responses of different species within a specific functional group (e.g. those insects that pollinate plants) towards changes or disturbances in the environment (Elmqvist *et al.* 2003; Ahern 2011). Redundancy refers to the variety of elements (e.g. species) fulfilling the same functions in a system, and when one of them is vulnerable to changes or disturbances the remaining elements can fulfil its function (Elmqvist *et al.* 2003). Ahern (2011) expands the concepts of response diversity and redundancy to be applied in the planning and design of urban bio-physical and social systems (green infrastructure planning). He uses examples of different elements of a 'green' urban stormwater management system such as permeable paving, bioswales and urban trees that intercept rainfall as different contributors to enhance the resilience capacity of the stormwater system by increasing the number of ecosystem services provided by the system (Ahern 2011). He also refers to the importance of social and economic diversity in cities showing a 'more complex response diversity by which they are better positioned to adapt to change and socio-economic disturbance' (Ahern 2011, p. 342).

Urban green infrastructure is internationally recognized as a concept that refers to the entire urban green network, including all natural, semi-natural and man-made green spaces and the connections between them at different spatial scales (Tzoulas *et al.* 2007; Pauleit *et al.* 2011). Expansion of the green infrastructure is regarded as one strategy to build resilience in cities (Meerow and Newell 2016). A green infrastructure approach in planning acknowledges that natural assets can fulfil the same functions (in the form of ecosystem services) as grey infrastructure such as transport and communication networks, water supply and wastewater systems (Pauleit *et al.* 2011). Green infrastructure planning includes five principles that should be taken into consideration in the management of the urban green infra-structure, namely (1) multifunctionality, (2) connectivity, (3) integration, (4) social inclusion and (5) following a long-term approach allowing adaptation through the involvement of different stakeholders (Ahern 2011; Pauleit *et al.* 2011).

The first principle of green infrastructure planning is that urban green spaces should be planned, designed and managed to be multifunctional, providing a variety of provisioning, supporting, regulating and cultural ecosystem services (Pauleit *et al.* 2011). As discussed earlier, cities have globally taken cognisance of the plight of biodiversity hotspots within urban areas. However, in cities that are not located in biodiversity hotspots, these usual conservation arguments do not hold and much emphasis has been placed on the importance of biodiversity to provide ecosystem services to argue for the development and management of green infrastructure (Roberts *et al.* 2012). Management of green infrastructure in African cities, therefore, considers the maintenance of ecosystem services as the most appropriate approach (Nkambule *et al.* 2016), as their value for poor urban residents is well documented (Roy *et al.* 2012).

Connectivity between the different parts of the green infrastructure (second principle) should be promoted at different scales, but also from different perspectives such as the ecological connectivity for species dispersal, the social connectivity for recreation and the physical connectivity for water flow and climate management

(Pauleit *et al.* 2011). It is generally accepted that enhancement and maintenance of biodiversity in cities is achieved by means of green infrastructure, including corridors and interconnectivity within the urban matrix (Hostetler *et al.* 2011). It has recently been shown that patch area and corridors have the strongest positive effects on biodiversity, complemented by vegetation structure (Beninde *et al.* 2015). However, some species are not dependant on the urban matrix and require different management actions. For example, movement of wetland passerines in cities suggests that patch management matters more than matrix management (Calder *et al.* 2015), due to their ability to fly and easily cross over buildings, roads and other obstructions. It is therefore clear that increasing both the area of habitat patches and creating a network of corridors is the most important strategy to maintain high levels of urban biodiversity for all species functional groups. Attempts should be made to prevent the green infrastructure from becoming too fragmented, resulting in potential extinction debts that have important consequences for urban biodiversity conservation and sustainable planning in African cities (Du Toit *et al.* 2016).

Integration between green infrastructure and grey infrastructure (third principle) such as buildings, roads and water systems will enhance the multiple functions that can be provided by each structure (Pauleit *et al.* 2011). This principle refers more specifically to the integration of the built landscape in the form of engineered (man-made) ecosystems such as green roofs, green walls and green stormwater management systems that 'mimic natural ecosystems to provide ecosystem services' (Oberndorfer *et al.* 2007, p. 831). By using various elements in the development of engineered ecosystems (e.g. green stormwater management system) fulfilling the same functions (redundancy and response diversity) could enhance resilience (Ahern 2011) as discussed earlier in this section. Although engineered ecosystems are increasingly dealt with in planning and management of African cities, this is often done in isolation of the overall green infrastructure, and therefore needs much research focus in the future.

The different types of green spaces in the green infrastructure and their interactions with grey infrastructure are managed by public, institutional and private owners. Thus, a participatory approach between various stakeholders (fourth principle of social inclusion) is needed to manage the urban green infrastructure (Pauleit *et al.* 2011). One example from Cape Town is the comparison of civic-led (local resident volunteers) and expert-led (ecologists) interventions in greening efforts using indigenous vegetation, concluding that both efforts were equally successful in terms of meeting conservation targets, but acknowledging the challenges of integrating civic-led interventions with biodiversity management (Anderson *et al.* 2014). An example from Durban is the involvement of various stakeholders in ecosystem-based adaptation (Roberts *et al.* 2012) that will be discussed in the next paragraph explaining the fifth principle of green infrastructure planning.

This final principle of ecosystem-based adaptation for green infrastructure planning calls for a strategic, long-term approach that considers the other four principles and acknowledges the importance of adaptation through the engagement of different stakeholders and the co-creation and sharing of knowledge based on urban biodiversity and ecosystem services (Pauleit *et al.* 2011). The city of Durban, South Africa can be used as an example of ecosystem-based adaptation as it is regarded as a global leader in climate adaptation (Roberts *et al.* 2011). Durban is also one of only seven cities from Africa that is included in the 100 resilient cities in the world by the Rockefeller Foundation (2017) based on how successful certain actions were taken to improve resilience focusing on health and well-being, economy and society, infrastructure and environment, and leadership and strategy. The Municipal Climate Protection Program of Durban includes three components, namely municipal adaptation (adaptation plans for key line functions such as water and health), community-based adaptation (to improve the adaptive capacities of local communities) and specific interventions in urban management addressing aspects such as urban heat islands, sea-level rise and stormwater runoff (Roberts *et al.* 2011). A good example of a long-term investment in Durban is the community reforestation projects that have positive socio-economic and biodiversity conservation impacts (Roberts *et al.* 2011).

In cities of developing African countries, green infrastructure planning enhancing urban resilience tend to be more often implemented in affluent suburbs than in the low income areas (Kuruneri-Chitepo and Shackleton 2011) following examples from the global north (Ziervogel *et al.* 2017). Urbanized areas of South Africa with greater per capita income and a higher proportion of green space are more likely to have higher woody species richness and more woody plants (McConnachie *et al.* 2008). This is a major shortcoming, as Mensah (2014) found that a well-managed system of green landscapes in resource-poor urban areas can generate net social benefits under a range of future climate change scenarios. Endeavours to establish green infrastructure in low income suburbs will create permanent reservoirs of biodiversity to sustain the relative ecosystem services, though budget and human resources constraints could represent serious impediments (Kuruneri-Chitepo and Shackleton 2011). Ziervogel *et al.* (2017, p. 124) also made the point that in cities with high inequality it is not only the 'hard infrastructure, technical engineering and ecosystem services' that need to be resilient but also the rights of urban citizens that are facing numerous risks on a daily basis. A challenge in addressing issues of equity and environmental justice in achieving urban resilience is to focus on the underlying politics of resilience by answering the questions of resilience 'For whom?', 'Resilience of what to what?', 'Short- or long-term resilience?', 'What are boundaries of the city that must be resilient?' and 'What is the goal of building resilience?' (Meerow and Newell 2016), all issues that are beyond the scope of this chapter.

Conclusions

Some major cities are located in the world's hotspots of biodiversity. Country legislation and international commitments (e.g. Aichi biodiversity targets) urge cities to address conservation priorities. These measures are dependent on appropriate ecological management practices, such as alien plant eradication, fire control and green space and infrastructure planning. Much of these actions are frowned upon by residents, developers and policymakers alike, especially as neighbouring cities might not be subjected to the same conservation priorities. People therefore feel burdened by conservation actions due to the shortage of land available for city development.

Management actions to safeguard biodiversity highlight the importance of providing ecosystem services through urban green spaces and infrastructures. These services benefit all levels of society and improve the quality of life and human well-being. However, important ecosystem services cannot always be linked to the critical biodiversity areas in systematic conservation plans. This can be overcome by arguing that all green space should be managed for ecosystem-based adaptation in response to future climate and environmental change scenarios by maintaining sufficient biodiversity to ensure redundancy and enhance urban resilience.

In African countries, the perceived value of ecosystem goods and services of public and private green space differ between socio-economic classes and cultural groups. For westernized residents, management of both public and private space is considered functional if it is of aesthetic value. In contrast, African culture applies indigenous knowledge systems to utilize the provisioning services of urban green space. These different perspectives are somehow conflicting, as in the way western culture manages their private gardens, biodiversity is enhanced, but in the way that African culture manages their public open spaces, biodiversity is also enhanced.

We propose that all these different urban green spaces need to be integrated into the urban green infrastructure that should be planned and co-managed by all stakeholders to be multifunctional, connected, integrated with the built environment and focusing on adaptation as a long-term planning and management strategy aiming towards building resilient cities. Moreover, by translating ecological knowledge into practice, we can create biodiverse liveable cities where both people and nature win.

References

Ahern J. (2011) From fail-safe to safe-to-fail: Sustainability and resilience in the new urban world. *Landscape and Urban Planning* 100: 341–343.

Alston K.P. and Richardson D.M. (2006) The roles of habitat features, disturbance, and distance from putative source populations in structuring alien plant invasions at the urban/wildland interface on the Cape Peninsula, South Africa. *Biological Conservation* 132(2): 183–198.

Anderson P.M.L., Avlonitis G. and Ernstson H. (2014) Ecological outcomes of civic and expert-led urban greening projects using indigenous plant species in Cape Town, South Africa. *Landscape and Urban Planning* 127: 104–113.

Aronson M.F.J., La Sorte F.A., Nilon C.H., Katti M., Goddard M.A., Lepczyk C.A., Warren P.S., Williams N.S.G., Cilliers S.S., Clarkson B., Dobbs C., Dolan R., Hedblomm M., Klotz S., Louwe Kooijmans J., Kühnn I., Macgregor-Fors I., Mcdonnell M.J., Mörtberg U., Pysek P., Siebert S.J., Sushinsky J., Werner P. and Winter M. (2014) A global analysis of the impacts of urbanization on bird and plant diversity reveals key anthropogenic drivers. *Proceedings of the Royal Society Biological Sciences*, 281: 1780.

Balmford A., Moore J.L., Brooks T., Burgess N., Hansen L.A., Williams P. and Rahbek C. (2001) Conservation conflicts across Africa. *Science* 291: 2616–2619.

Beninde J., Veith M. and Hochkirch A. (2015) Biodiversity in cities needs space: A meta-analysis of factors determining intra-urban biodiversity variation. *Ecology Letters* 18(6): 581–592.

Bigirimana J., Bogaert J., De Cannière C., Bigendako M. J. and Parmentier I. (2012) Domestic garden plant diversity in Bujumbura, Burundi: Role of the socio-economical status of the neighborhood and alien species invasion risk. *Landscape and Urban Planning* 107(2): 118–126.

Boon R., Cockburn J., Douwes E., Govender N., Ground L., Mclean C., Roberts D., Rouget M. and Slotow R. (2016) Managing a threatened savanna ecosystem (KwaZulu-Natal Sandstone Sourveld) in an urban biodiversity hotspot: Durban, South Africa. *Bothalia* 46(2): a2112.

Buizer M., Elands B. and Vierikko K. (2016) Governing cities reflexively: The biocultural diversity concept as an alternative to ecosystem services. *Environmental Science & Policy* 62: 7–13.

Calder J.L., Cumming G.S., Maciejewski K. and Oschadleus H.D. (2015) Urban land use does not limit weaver bird movements between wetlands in Cape Town, South Africa. *Biological Conservation* 187: 230–239.

Cilliers S.S. (2010) Social aspects of urban biodiversity: An overview. In Müller, N., Werner, P. and Kelcey, J. (eds), *Urban Biodiversity & Design: Implementing the Convention on Biological Diversity in Towns and Cities*, pp. 81–100. Wiley-Blackwell, Oxford.

Cilliers S.S. and Siebert S.J. (2012) Urban ecology in Cape Town: South African comparisons and reflections. *Ecology and Society* 17(3): 33. Available at http://dx.doi.org/10.5751/ES-05146-170333

Cilliers S.S., Bouwman H. and Drewes E. (2009) Comparative urban ecological research in developing countries. In McDonnell, M.J., Hahs, A.K. and Breuste, J.H. (eds), *Ecology of Cities and Towns*, pp. 90–111. Cambridge University Press, Cambridge.

Cilliers S.S., Cilliers J., Lubbe R. and Siebert S.J. (2013) Ecosystem services of urban green spaces in African countries: Perspectives and challenges. *Urban Ecosystems* 16(4): 681–702.

Cocks M. (2006) Biocultural diversity: Moving beyond the realm of 'indigenous' and 'local' people. *Human Ecology* 34(2): 185–200.

Davids R., Rouget M., Boon R. and Roberts D. (2016) Identifying ecosystem service hotspots for environmental management in Durban, South Africa. *Bothalia* 46(2): a2118.

Davoren, E. (2016) Plant diversity patterns of domestic gardens in five settlements of South Africa. PhD thesis, North-West University, Potchefstroom, South Africa.

Davoren E., Siebert S.J., Cilliers S.S. and Du Toit M.J. (2016) Influence of socioeconomic status on design of Batswana home gardens and associated plant diversity patterns in northern South Africa. *Landscape and Ecological Engineering* 12(1): 129–139.

Driver A., Cowling R.M. and Maze K. (2003) Planning for Living Landscapes: Perspectives and lessons from South Africa. Available at www.cepf.net/documents/living.landscapes.pdf (accessed 29 March 2017).

Du Toit M.J., Kotze D.J. and Cilliers S.S. 2016. Landscape history, time lags and drivers of change: Urban natural grassland remnants in Potchefstroom, South Africa. *Landscape Ecology* 31(9): 2133–2150.

Egoh B., Reyers B., Rouget M., Bode M. and Richardson D.M. (2009) Spatial congruence between biodiversity and ecosystem services in South Africa. *Biological Conservation* 142: 553–562.

Elmqvist T. (2014) Urban resilience thinking. *Solutions* 5(5): 26–30.

Elmqvist T., Folke C., Nyström M., Peterson G., Bengtsson J., Walker B. and Norberg J. (2003). Response diversity, ecosystem change and resilience. *Frontiers in Ecology and the Environment* 1(9): 488–494.

Fanan U., Dlama K.I. and Oluseyi I.O. (2011) Urban expansion and vegetal cover loss in and around Nigeria's Federal capital city. *Journal of Ecology and the Natural Environment* 3(1): 1–10.

Farinha-Marques P., Lameiras J.M., Fernandes C., Silva S. and Guilherme F. (2011) Urban biodiversity: A review of current concepts and contributions to multidisciplinary approaches. *The European Journal of Social Science Research* 24(3): 2472–71.

Gaertner M., Larson B.M.H., Irlich U.M., Holmes P.M., Stafford L., Van Wilgen B.W. and Richardson D.M. (2016) Managing invasive species in cities: A framework from Cape Town, South Africa. *Landscape and Urban Planning* 151: 1–9.

Gill N., Waitt G. and Head L. (2009) Local engagements with urban bushland: Moving beyond bounded practice for urban biodiversity management. *Landscape and Urban Planning* 93: 184–193.

Goddard M., Dougill A.J. and Benton T.G. (2010) Scaling up from gardens: Biodiversity conservation in urban environments. *Trends in Ecology and Evolution* 25: 90–98.

Gong C., Chen J. and Yu S. (2013) Biotic homogenization and differentiation of the flora in artificial and near-natural habitats across urban green spaces. *Landscape and Urban Planning* 120: 158–169.

Graham M. and Ernstson H. (2012) Comanagement at the fringes: examining stakeholder perspectives at Macassar Dunes, Cape Town, South Africa—at the intersection of high biodiversity, urban poverty, and inequality. *Ecology and Society* 17(3): 34. Available at http://dx.doi.org/10.5751/ES-04887-170334.

Ground L.E., Slotow R. and Ray-Mukherjee J. (2016) The value of urban and peri-urban conservation efforts within a global biodiversity hotspot. *Bothalia* 46(2): a2106.

Hahs A. K. and McDonnell M. J. (2014) Extinction debt of cities and ways to minimise their realisation: a focus on Melbourne. *Ecological Management & Restoration* 15(2): 102–110.

Hoffman T. S. and O'Riain M. J. (2012) Monkey management: Using spatial ecology to understand the extent and severity of human–baboon conflict in the Cape Peninsula, South Africa. *Ecology and Society* 17(3): 1-16. Available at http://dx.doi.org/10.5751/ES-04882-170313.

Holmes P.M., Rebelo A.G., Dorse C. and Wood J. (2012) Can Cape Town's unique biodiversity be saved? Balancing conservation imperatives and development needs. *Ecology and Society* 17(2): 28. Available at https://doi.org/10.5751/ES-04552-170228.

Hope D., Gries C., Zhu W., Fagan W.F., Redman C.L., Grimm N.B., Nelson A.L., Martin C. and Kinzig A. (2003) Socioeconomics drive urban plant diversity. *Proceedings of the National Academy of Sciences of the United States of America* 100: 8788-8792.

Hostetler M., Allen W. and Meurk C. (2011) Conserving urban biodiversity? Creating green infrastructure is only the first step. *Landscape and Urban Planning* 100(4): 369–371.

Idohou R., Fandohan B., Salako V.K., Kassa B., Gbèdomon R.C., Yédomonhan H., Glèlè Kakaï R.L. and Assogbadjo A.E. (2014) Biodiversity conservation in home gardens: Traditional knowledge, use patterns and implications for management. *International Journal of Biodiversity Science, Ecosystem Services & Management* 10(2): 89–100.

Izidine S.A., Siebert S.J., Van Wyk A.E. and Zobolo A.M. (2008) Taboo and political authority in conservation policy: A case study of the Licuáti Forest in Maputaland, Mozambique. *Journal for the Study of Religion, Nature, and Culture* 2(3): 373–390.

Kuruneri-Chitepo C. and Shackleton C.M. (2011) The distribution, abundance and composition of street trees in selected towns of the Eastern Cape, South Africa. *Urban Forestry & Urban Greening* 10(3): 247–254.

Lemanski C. (2004) A new apartheid? The spatial implications of fear of crime in Cape Town, South Africa. *Environment and Urbanization* 16(2): 101–112.

Loram A., Warren P., Thompson K. and Gaston K. (2011) Urban domestic gardens: The effects of human interventions on garden composition. *Environmental Management* 48: 808–824.

Lubbe C.S., Siebert S.J. and Cilliers S.S. (2010) Political legacy of South Africa affects the plant diversity patterns of urban domestic gardens along a socio-economic gradient. *Scientific Research and Essays* 5(19): 2900–2910.

Lyytimäki J. and Sipilä M. (2009) Hopping on one leg: The challenge of ecosystem dis-services for urban green management. *Urban Forestry & Urban Greening* 8(4): 309–315.

Matsiliza B. and Barker N.P. (2001) A preliminary survey of plants used in traditional medicine in the Grahamstown area. *South African Journal of Botany* 67: 177–182.

McConnachie M.M., Shackleton C.M. and McGregor G.K. (2008). The extent of public green space and alien plant species in 10 small towns of the Sub-Tropical Thicket Biome, South Africa. *Urban Forestry & Urban Greening* 7(1): 1–13.

McDonnell M. and Hahs A. (2014) Four Ways to Reduce the Loss of Native Plants and Animals from Our Cities and Towns, The Nature of Cities. Available at www.thenatureofcities.com/2014/04/14/four-ways-to-reduce-the-loss-of-native-plants-and-animals-from-our-cities-and-towns/ (accessed 6 August 2017).

McDonnell M. J. (2007) Restoring and managing biodiversity in an urbanizing world filled with tensions. *Ecological Management & Restoration* 8(2): 83–84.

Meerow S. and Newell J.P. (2016) Urban resilience for whom, what, when, where and why? *Urban Geography*. Available at http://dx.doi.org/10.1080/02723638.2016.1206395.

Mensah C.A. (2014) Urban green spaces in Africa: Nature and challenges. *International Journal of Ecosystem* 4(1): 1–11.

Molebatsi L.Y., Siebert S.J., Cilliers S.S., Lubbe C.S. and Davoren E. (2010) The Tswana tshimo: A homegarden system of useful plants with a specific layout and function. *African Journal of Agricultural Research* 5(21): 2952–2963.

Mpofu T.P.Z. (2013) Environmental challenges of urbanisation: A case study for open green space management. *Research Journal of Agricultural and Environmental Management* 2(4): 105–110.

Muderere T. 2011. Natural coexistence or confinement: Challenges in integrating birdlife concerns into urban planning and design for Zimbabwe. *Journal of Sustainable Development in Africa* 13(1): 162–183.

Nilon C. (2011) Urban biodiversity and the importance of management and conservation. *Landscape Ecology and Engineering* 7: 45–52.

Nkambule S.S., Buthelezi H.Z. and Munien S. (2016) Opportunities and constraints for community-based conservation: The case of the KwaZulu-Natal Sandstone Sourveld grassland, South Africa. *Bothalia* 46(2): a2120.

Oberndorfer E., Lundholm J., Bass B., Coffman R.R., Doshi H., Dunnett N., Gaffin S., Köhler M., Liu K.K.Y. and Rowe B. (2007) Green roofs as urban ecosystems: Ecological structures, functions, and services. *BioScience* 57(10): 823–833.

O'Farrell P. J., Anderson P.M.L., Le Maitre D.C. and Holmes P.M. (2012) Insights and opportunities offered by a rapid ecosystem service assessment in promoting a conservation agenda in an urban biodiversity hotspot. *Ecology and Society* 17(3): 27. Available at https://doi.org/10.5751/ES-04886-170327.

Ossola A., Hahs A.K., Nash M.A. and Livesley S.J. (2016) Habitat complexity enhances comminution and decomposition processes in urban ecosystems. *Ecosystems* 19(5): 927–941.

Pauleit S., Liu L., Ahern J. and Kazmierczak A. (2011) Multifunctional green infrastructure to promote ecological services in the city. In Niemela J. (ed.), *Urban Ecology: Patterns, Processes and Applications*, pp. 272–285. Oxford University Press, New York.

Pauw A. and Louw K. (2012) Urbanization drives a reduction in functional diversity in a guild of nectar-feeding birds. *Ecology and Society* 17(2): 27. Available at http://dx.doi.org/10.5751/ES-04758-170227

Pearson R.G. (2016) Reasons to conserve nature. *Trends in Ecology and Evolution* 31(5): 366–371.

Pryke J.S. and Samways M.J. (2009) Recovery of invertebrate diversity in a rehabilitated city landscape mosaic in the heart of a biodiversity hotspot. *Landscape and Urban Planning* 93(1): 54–62.

Rebelo A.G., Holmes P.M., Dorse C. and Wood J. (2011) Impacts of urbanization in a biodiversity hotspot: Conservation challenges in Metropolitan Cape Town. *South African Journal of Botany* 77(1): 20–35.

Reuther S. and Dewar N. (2005) Competition for the use of public open space in low-income urban areas: The economic potential of urban gardening in Khayelitsha, Cape Town. *Development Southern Africa* 23(1): 97–122.

Roberts D., Boon R., Diederichs N., Douwes E., Govender N., McInnes A., McLean, C., O'Donoghue, S. and Spires M. (2012) Exploring ecosystem-based adaptation in Durban, South Africa: 'learning-by-doing' at the local government coal face. *Environment and Urbanization* 24(1): 167–195.

Rockefeller Foundation. (2017) 100-Resilient Cities. Available at www.100resilientcities.org (accessed 21 April 2017).

Roy S., Byrne J. and Pickering C. (2012) A systematic quantitative review of urban tree benefits, costs, and assessment methods across cities in different climatic zones. *Urban Forestry & Urban Greening* 11(4): 351–363.

Shackleton C.M. (2006) Urban forestry: A Cinderella science in South Africa? *Southern African Forestry Journal* 208: 1–4.

Sperling C.D. and Lortie C.J. (2010) The importance of urban backgardens on plant and invertebrate recruitment: A field microcosm experiment. *Urban Ecosystems* 13: 223–235.

Swanwick C., Dunnett N. and Woolley H. (2003) Nature, role and value of green space in towns and cities: An overview. *Built Environment* 29(2): 94–106.

Tennent J., Downs C.T. and Bodasing M. (2009) Management recommendations for feral cat (*Felis catus*) populations within an urban conservancy in KwaZulu-Natal, South Africa. *South African Journal of Wildlife Research* 39(2): 137–142.

Tzoulas K., Korpela K., Venn S., Yli-Pelkonen V., Kazmierczak A. and Niemelä J. (2007) Promoting ecosystem and human health in urban areas using green infrastructure: A literature review. *Landscape and Urban Planning* 81: 167–178.

Van Wilgen B. W., Forsyth G.G. and Prins P. (2012) The management of fire-adapted ecosystems in an urban setting: the case of Table Mountain National Park, South Africa. *Ecology and Society* 17(1): 8. Available at http://dx.doi.org/10.5751/ES-04526-170108

Von Döhren P. and Haase D. (2015) Ecosystem disservices research: A review of the state of the art with a focus on cities. *Ecological Indicators* 52: 4904–4997.

Walker B. and Salt D. (2006). *Resilience Thinking*. Island Press, Washington, DC.

Whitmore C., Crouch T.E. and Slotow R.H. (2002) Conservation of biodiversity in urban environments: invertebrates on structurally enhanced road islands. *African Entomology* 10(1): 113–126.

Ziervogel G., Pelling M., Cartwright A., Chu E., Desphande T., Harris L., Hyams K., Kaunda J., Klaus B., Michael K., Pasquini L., Pharoah R., Rodina L., Scott D. and Zweig P. (2017) Inserting rights and justice into urban resilience: A focus on everyday risk. *Environment & Urbanization* 29(1): 123–138.

12

URBAN GREEN INFRASTRUCTURES AND ECOLOGICAL NETWORKS FOR URBAN BIODIVERSITY CONSERVATION

Emily S. Minor, Elsa C. Anderson, J. Amy Belaire, Megan Garfinkel and Alexis Dyan Smith

Introduction

As urban areas develop, native habitats are often degraded, fragmented, and replaced by impervious surfaces. These changes can lead to a suite of environmental problems, including the urban heat island effect, flooding, water and air pollution, and reduced biodiversity (Grimm *et al.* 2008). Green infrastructure has been proposed as a solution to these problems (Benedict and McMahon 2006). In this chapter, we focus on the benefits of green infrastructure networks for biodiversity, and suggest strategies for increasing the extent, performance, and connectivity of these networks to promote urban habitats and native species.

Although definitions of *green infrastructure* can vary across spatial scales, the term itself suggests that nature—just like our built infrastructure—is critical for society. At the local scale, the term is often applied to environmentally friendly stormwater management systems such as bioswales or permeable pavement. At the larger landscape scale, green infrastructure describes interconnected green spaces and strategically planned networks of natural and semi-natural areas across a region. This definition incorporates green and blue spaces in rural, urban, terrestrial, coastal, and marine systems (European Commission 2013). Importantly, this second definition also sees green infrastructure as multifunctional spaces that can benefit humans, wildlife, and the larger environment.

Over the last few decades, many cities and metropolitan areas have developed large-scale green infrastructure plans. These plans often aim to protect biodiversity and provide a wide array of other benefits by preserving and restoring large core habitats. The benefits, referred to as *ecosystem services* (Daily 1997), include stormwater management, urban heat island mitigation, and recreation opportunities. Another common goal of these plans is maintaining or increasing *connectivity* between the core habitats. Connectivity is seen as desirable ecologically because interconnected habitats are more likely to maintain natural communities and ecological processes (Fischer and Lindenmayer 2007). Connectivity may also be socially desirable, as linked green spaces can facilitate walking or cycling for recreation and transportation.

While green infrastructure can be structurally connected—for example, as contiguous *greenbelts* around some cities—structural connectivity may be difficult to attain in cities that have been built up for decades or centuries. On the other hand, functional connectivity, which refers to the movement of organisms and continuity of ecological processes (Tischendorf and Fahrig 2000), may be more achievable in most urban areas. Functional connectivity can be encouraged in landscapes that are not structurally connected by considering the area between large green spaces (called *the matrix*).

Cities typically include many small green spaces within the matrix, including home gardens, community gardens, and vacant lots. These small green spaces, in combination with other larger spaces such as cemeteries and golf courses, are often overlooked ecologically and have not typically been considered as green infrastructure in the same way as designated natural areas or forest preserves. But when embedded in larger green infrastructure networks, they can serve to increase functional connectivity by facilitating species movement between core habitats. Furthermore, these unconventional types of green infrastructure represent an opportunity to provide ecological and social benefits in built-out communities that have limited ability to create new parks.

The value of parks and preserves for urban biodiversity conservation has been written about extensively elsewhere. In the rest of this chapter, we focus on the role of smaller and less-conventional green spaces. We discuss four abundant types of green spaces that are found in cities around the world—home gardens, urban agriculture patches, vacant lots, and cemeteries—and the role they can play in urban green infrastructure networks. These smaller spaces, in the aggregate, are often underestimated in terms of their potential to contribute to urban green infrastructure networks and conservation of biodiversity. After introducing each type of green infrastructure and its potential benefits for biodiversity, we end the chapter with a section that considers them in combination with other green spaces, as part of the urban green infrastructure network. We aim to make broader conclusions and recommendations but draw many examples from our experience with green infrastructure in the Chicago metropolitan area (Illinois, USA).

The role of the matrix in urban green infrastructure networks

Vacant lots

Urban vacant lots—sometimes called wastelands, greenfields, or abandoned/derelict land—can serve as a multifunctional resource in cities (Anderson and Minor 2017). Vacant land has many definitions but here we refer to previously developed residential or commercial parcels that have fallen out of use and no longer contain buildings. By this definition, the City of Chicago owns >13,000 vacant lots and contains at least another 10,000 bank-owned lots. Most large cities currently have between 12.5 percent and 15 percent vacancy, but the American Rustbelt—comprised of Midwestern cities with a historically manufacturing-based economy—has seen a sizable increase in vacancy due to loss of industry and displacement of the working class. The most notorious example of contemporary vacancy in the United States is Detroit, Michigan, which has seen vacancy rates as high as 35 percent in the last decade (Gallagher 2010). Overall, this represents 100s to 1000s of hectares of land per city that could provide ecosystem services and wildlife habitat.

The ecology and ecosystem services in vacant lots are dependent on their history and management. Occasionally ornamental plants or garden features persist, but they are unlikely to be maintained. In fact, the level of environmental management is a critical difference between vacant lots and other types of green infrastructure. Management such as mowing or spraying weeds can be contextualized as an ecological disturbance where humans reset or manipulate the plant community. Certain plant species are better competitors under different conditions, and management actively changes these conditions over time. There is also some evidence that vacant lots can facilitate persistence of plant populations in changing climates, as dispersed seeds are more likely to survive in a vacant lot than in other heavily managed green spaces (Renton *et al.* 2014).

While individual lots are subject to various management activities, as a whole, vacant lots provide relatively consistent habitat across the city. People perceive many plants in vacant lots as weeds, but these communities support surprising wildlife diversity. For example, vacant lots are home to a large diversity of insects and other arthropods, and the composition of these communities differs significantly from those in other urban habitats (Gardiner *et al.* 2013). Vacant lots also provide habitat for small mammals (Magle *et al.* 2010), and various rare species have been found in vacant lots as well (Vessel and Wong 1987). Furthermore, trees in vacant lots may provide resources for arboreal species. These trees are more likely than street trees to have dead or decayed branches, which can be beneficial for birds such as woodpeckers that prefer softened wood for finding food and excavating nesting cavities. Finally, vacant lots can serve as *stepping-stones* for species moving between larger green spaces, mitigating the risks of isolating populations in preserves (i.e. *ecological traps*).

Some simple steps can improve environmental quality and biodiversity of vacant lots. The City of Chicago spends over $1.2 million annually on controlling weeds in vacant lots, and mowing and lawn care are significant sources of fossil fuel consumption and pollution (Robbins *et al.* 2001). Planting lower-maintenance trees or flowers in these spaces could reduce the economic and ecological costs of mowing, while increasing air quality and providing habitat for local species of wildlife to traverse the urban matrix.

A primary challenge with vacant lots is that they are disproportionately located in low-income and minority neighborhoods (Kremer *et al.* 2013). Environmental remediation and improvements create a risk of eco-gentrification (Dooling 2009), which is an environmental justice paradox in which the deliberate improvement of environmental quality draws in a higher economic class. Often the original residents—those whom the project was supposed to help—cannot afford the increased cost of living and are forced into areas of poorer environmental quality.

In light of this social issue, there are two primary mechanisms by which green infrastructure improvement in vacant lots can be socially and ecologically success-ful. Green infrastructure that is designed with input of community leaders and that engages and educates local residents is considered the gold standard (Pediaditi et al. 2010). These types of projects might include development of community gardens, nature play areas, or surrogate yards/parks where local residents can inter-act with and learn about nature in their own neighborhood. While these projects may or may not increase biodiversity on their own, increasing public awareness of environmental issues could potentially have trickle-down effects by building environmentally friendly behaviors and instilling an intrinsic or economic value for urban biodiversity.

The alternative option is to expand green infrastructure benefits in these areas in ways that are almost unnoticeable, yet still beneficial. This is referred to as the *just green enough* strategy for vacant lot greening and management (Curran and Hamilton 2012). For example, spreading seeds of robust and inexpensive native wildflowers might result in a plant community that is higher quality for wildlife without being visibly different from a weedy lot to an untrained eye. If these types of improvements are relatively unnoticeable, they can increase biodiversity and ecosystem service provision without incentivizing gentrification.

Home gardens

Private residential yards and gardens ("home gardens" from here forward) com-prise a large portion of the green space in many cities. Unlike many other kinds of green infrastructure, which are either widely spaced (e.g., cemeteries) or clustered in certain areas of a city (e.g., vacant lots), home gardens are more evenly dis-tributed and often structurally connected to each other. These small green spaces are the primary place where many humans come into contact with the natural world. They also provide numerous ecosystem services to urban residents, includ-ing absorbing stormwater, reducing cooling costs in the summer, and providing a place to relax and rejuvenate. And finally, they have the potential to contribute substantially to conservation of biodiversity (Davies et al. 2009).

Although lawns comprise a dominant vegetation type around some homes, particularly in the former British Empire, people's preferences for garden plants are very diverse (Kendal et al. 2012). As a result, many residential areas have sur-prisingly high plant diversity. Studies in Chicago (unpublished data), Melbourne (Threlfall et al. 2016), and New Zealand (van Heezik et al. 2013) each revealed over 500 plant species while a study in the UK found over 1000 plant species in home gardens (Loram et al. 2008).

In addition to a diversity of plants, many home gardens contain other resources for wildlife such as bird feeders and baths, nesting and hibernating substrates, and outdoor pet food (e.g. Belaire *et al.* 2014). Those resources can contribute to the diversity of wildlife species. For example, in Chicago, migratory birds were observed in neighborhoods with more wildlife-friendly features in yards (Belaire *et al.* 2014). That same study showed that local vegetation in home gardens was a better predictor of the bird community in residential neighborhoods than larger-scale vegetation features. Home gardens may be especially good habitat for smaller animals such as invertebrates. A study of home gardens in Sheffield (UK) identified over 350 invertebrate species (Smith *et al.* 2006). In Norman, Oklahoma, 32 species of land snails were found in home gardens (Bergey and Figueroa 2016). Residential neighborhoods also support a diversity of pollinators (Lowenstein *et al.* 2014). In each case, the number of invertebrate species was related to management (such as watering or use of pesticides), vegetation, or resources in home gardens. Even lawns, if left to grow some flowering weeds, can contribute to wildlife biodiversity (Larson *et al.* 2014).

Although biodiversity can be high in home gardens, many plants in them are non-native. The impact of exotic vegetation on wildlife remains an area of debate but generally it is thought that native plant species promote higher diversity of native wildlife. This may be particularly true for insects, which often have specific plant hosts in certain life stages, but the effect can also trickle up to higher trophic levels (Burghardt and Tallamy 2013). One way to fulfill the full ecological potential of home gardens might be to increase the number of native plant species and other wildlife resources they contain.

Options for incentivizing wildlife-friendly gardening fall into two broad categories: (i) top-down, financial incentives or regulation; and (ii) bottom-up, community-led initiatives (Goddard *et al.* 2009). Formal top-down institutions such as homeowners associations (HOAs) can be effective vehicles for converting entire neighborhoods into wildlife-friendly zones. Research in Phoenix, Arizona demonstrated that neighborhoods with HOAs had significantly greater bird and plant diversity than neighborhoods without (Lerman *et al.* 2012). Some cities and non-governmental agencies (NGOs) offer optional incentives or rebates for gardeners who plant native plants or implement other environmentally friendly management practices. One such NGO is the National Wildlife Federation (NWF), whose Certified Wildlife Habitat Program has certified over 200,000 wildlife habitats to date. NWF's program focuses on five key garden components: food, water, shelter, places to raise young, and sustainable gardening practices. Researchers found that home gardens certified by this program provided more abundant and higher quality wildlife habitat relative to non-certified gardens (Widows and Drake 2014). Finally, social norms and networks can also be effective motivators for design and management of home gardens. A citizen science program called Habitat Network, a partnership between The Nature Conservancy and Cornell Lab of Ornithology, attempts to use social networks to motivate users to integrate wildlife habitat elements into backyards, parks, and other green spaces. In the Habitat Network website, users examine their landscape with aerial imagery and identify partial corridors or gaps in connectivity that could be filled with site-scale enhancements. Because so much land in cities is privately owned, finding ways like these to engage private landowners in enhancing biodiversity on their land will be crucial to conservation and urban sustainability.

Urban agriculture

Urban agriculture is a rapidly expanding source of green space in many cities (Tornaghi 2014). Although some definitions of urban agriculture include cultivation of ornamental or non-food crop species, for this chapter we define urban agriculture as food crop and livestock production for use within the urban area where they were produced. Urban agriculture may include residential food gardens, commercial or educational urban farms, and community food gardens (including both school and neighborhood gardens).

While globally most crops are grown in the ground, urban agriculture is unique in that it may also be found on rooftops or in raised beds or containers. Rooftop gardens allow for increased green space in densely built areas within cities, and raised bed or container gardens allow for cultivation without having to remove underlying impermeable surfaces. Both increase green infrastructure in locations where other types of green spaces would be impossible to introduce without major construction and planning.

Within cities, the extent and distribution of urban agriculture is variable. Urban agriculture is often distributed unevenly across space, socio-economic gradients, and cultural lines. For instance, in Chicago, approximately 26 ha is used for some form of urban agriculture and community gardens are concentrated on the economically disadvantaged south and west sides (Taylor and Lovell 2012). Furthermore, neighborhoods with high concentrations of Chinese-origin, Eastern- or Southern-European, and Polish residents seem to be hot spots for home food gardens. In Philadelphia, Pennsylvania (USA), community gardens are also concentrated in low-income neighborhoods, but they are distributed even more unevenly than in Chicago, with most located in very poor neighborhoods (Kremer and DeLiberty 2011). Similarly, in developing countries, the poorest urban households often show higher percentages of agricultural activity (Thebo et al. 2014). This uneven distribution of urban agriculture may, in some cases, simply be correlated with availability of land, or may be directly influenced by the efforts of NGOs to increase local food availability or the cultural values and norms of the residents.

Because of the large variety of urban agriculture types, and the fact that most plots are managed by different people or groups in different ways, urban agriculture tends to be heterogeneous in both structural and plant diversity. Most urban gardens and farms grow a large variety of crops (e.g. Bernholt et al. 2009), especially when compared to large-scale rural agriculture. While the reasons for this may be social or economic, the resulting ecological effect is increased diversity of micro-habitats for wildlife. Fruit trees, climbing vines, and low row crops all contribute to structural diversity, and growing a combination of these crop types within a garden supports wildlife across a variety of taxa. A study in Niger found that the ethnicity of the gardener may determine both the number and type of species grown within the urban garden (Bernholt *et al.* 2009). Community gardens in particular may have very high levels of vegetative biodiversity (Lin *et al.* 2015), because different community members are free to grow crops according to their own tastes or interests.

All agricultural systems support agricultural pests, but urban agriculture tends to also support healthy populations of *natural enemy* arthropods as well (Lowenstein and Minor, in review). Urban agriculture may also provide habitat for a variety of pollinators (e.g. Matteson *et al.* 2008), and urban apiculture directly provides pollination services to nearby areas. Unfortunately, although there is some documentation of insect diversity in urban agriculture (likely because of the obvious services and disservices that insects provide in agriculture), specific information on other taxa is currently lacking. We do know, however, that in many cities urban agriculture tends to use more wildlife friendly practices than large-scale rural agriculture. Many urban farms use organic practices, and many community gardens restrict the use of chemicals such as pesticides and fertilizers (Brown and Jameton 2000). Moreover, community and school gardens often have plenty of participants willing to provide manual labor such as hand weeding and pest removal, which decreases the need for chemical inputs.

Future improvements to urban agriculture that benefit biodiversity can come from both top-down and bottom-up sources. While chemical fertilizers and pesticides are not often used in large quantities in most urban agriculture, the effect of any chemical runoff is intensified because there are generally no buffers between the gardens and neighboring green spaces or development. Top-down regulation of chemical inputs could increase the value of urban agricultural sites as habitat for wildlife, especially in home gardens where there is less accountability. Bottom-up changes in management and crop selection could also increase the value of urban agriculture habitat. For instance, where possible urban gardeners and farmers could increase the structural diversity of crops to provide more micro-habitats for wildlife. At the same time, urban gardeners must ensure that these sites do not become reservoirs only for common urban pest species, such as rats (*Rattus* species), raccoons (*Procyon lotor*), house sparrows (*Passer domesticus*), or cabbage white butterflies (*Pieris rapae*). Rats and raccoons, in particular, are attracted to easy food sources, and may actually harm other native wildlife through either competition or predation. To support native biodiversity within cities, urban gardeners should consider methods of controlling these urban-adapted pest species for the sake of other native wildlife.

Cemeteries

In some cities, cemeteries are among the oldest and most established components of urban green infrastructure. The *rural cemetery* movement in the United States was sparked in 1831 by the establishment of Mount Auburn, a 29 ha landscaped cemetery outside of Boston, Massachusetts (French 1974). At a time when cities were rapidly industrializing, these cemeteries (more accurately called *garden cemeteries*) were intended not only as a place to bury the dead, but also as a bucolic respite to those living in cities. Originally located outside of a city's limits, many large cemeteries remained even as the city expanded (e.g. Pattison 1955). Cities that developed during this garden cemetery movement tend to have similar patterns in extent and distribution of cemeteries. Today, the remaining cemeteries offer refuges to city-dwelling humans and wildlife alike (Barrett and Barrett 2001).

Garden cemeteries are the predecessors to the first public parks in the United States (Shelton 2008). Like public parks, cemeteries confer multiple green infra-structure benefits including air filtration, retention of contaminants and sediment, and mitigation of urban heat island effects. Cemeteries contain some of the oldest and largest trees found in cities. In Dayton, Ohio for example, a champion 27 m tall sassafras tree (*Sassafras albidum*) grows in the Woodland Cemetery (Barrett and Barrett 2001). Some cemeteries, such as Chicago's Graceland Cemetery, are cer-tified arboretums. Graceland manages over 2,000 trees, 50 of which are unique species mapped for visitors who wish to take a tree tour.

Cemeteries also offer unique biodiversity benefits. Many cemeteries have walls and gates that are closed from evening to early morning, when avian activity is at its highest (Lussenhop 1977). This allows breeding birds to sing for mates and nocturnal species to hunt with reduced human disturbance. Insectivorous birds will sometimes use headstones and other monuments as perches while hunting for insects. There are perhaps fewer social and aesthetic constraints on dying trees in cemeteries; a snag in a cemetery may go unnoticed, whereas a snag in front of someone's home raises concerns about lowered property values, or a tree limb fall-ing on a person or a parked car. These large old trees provide crucial habitat for many species (Cockle *et al.* 2011). For lichen species, cemeteries provide a broad range of substrates in the form of trees, headstones, and monuments. In places such as Illinois where exposed stone is rare, headstones can be important substrates for saxicolous lichens (Wachholder *et al.* 2004). Many cemeteries are large in area and experience low levels of human disturbance, attributes that allow them to serve as habitat to larger mammals and birds, such as coyotes and owls.

It should be noted, however, that some aspects of cemeteries can detract from their value as green spaces. To maintain their appearance, cemeteries are often landscaped using chemical inputs and lawnmowers. The chemicals used to preserve the buried are toxic, and caskets are often placed in concrete vaults (Basmajian and Coutts 2010). Tree species found in cemeteries are not always native and some, such as Norway Maples (*Acer platanoides*), are even invasive. Horizontally oriented black headstones may function as an ecological trap for dragonflies, because they polarize light in a similar way to the surface of water, where dragonflies lay their eggs (Horváth *et al.* 2007). These examples demon-strate that although cemeteries play a valuable role in urban green infrastructure networks, there is certainly room for improvement.

Newer cities that did not grow during the garden cemetery movement often overlook planning for the dead (Basmajian and Coutts 2010). The certainty of human mortality and the finite nature of land demand that future planning for the dead should take place, and there is potential for such planning to be incorporated in green infrastructure networks. Moreover, revenues from "green" burials, which forego flesh preservatives and concrete, on a small parcel of land could be used to preserve or restore larger natural areas (Basmajian and Coutts 2010).

Networks of green infrastructure

Ecological land-use complementation is the idea that land uses in urban areas could synergistically interact to support biodiversity and realize emergent ecological functions when clustered together in different combinations (Colding 2007). Vacant lots, home gardens, urban agriculture, and cemeteries, in conjunction with remnant habitats, parks, and preserves, have different spatial distributions, offer different types of habitat, and make unique contributions to urban biodiversity (Table 12.1, Figure 12.1). But together, in a network of interconnected green infrastructure, they can be greater than the sum of their parts. Collectively, the various land uses within the city form a mosaic, and more green infrastructure within that mosaic can enhance functional connectivity and support species movement across the landscape.

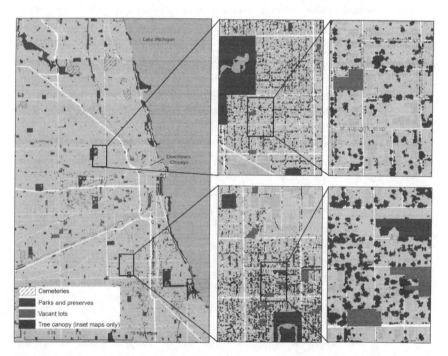

FIGURE 12.1   Green infrastructure networks at different spatial scales across Chicago. At broad spatial scales (larger map), cemeteries, parks and preserves, and vacant lots appear to dominate the green infrastructure network. At smaller spatial scales (inset maps), tree canopy and other vegetation in home gardens and residential neighborhoods provides habitat connectivity between the larger green spaces. (Source: Emily Minor, unpublished.)

Strategies that address multiple spatial scales are necessary to enhance green infrastructure networks and functional connectivity. At regional scales, government agencies and planning organizations can dedicate open space in locations that enhance connectivity across the landscape, including riparian corridors, habitat remnants, and small stepping-stone patches that facilitate movement. For example, in Austin, Texas (USA), government partners worked to preserve over 16,000 hectares of land through outright purchases and conservation easements, with the goal of protecting high quality habitat for biodiversity and water quality in face of rapid urbanization in the region. The preserved habitats are distributed across much of the county and include remnant woodland patches and an 11 km long riparian corridor in the city limits. Collectively, they harbor eight federally listed endangered species and more than 20 species of conservation concern.

At smaller spatial scales, city governments can increase habitat connectivity with strategies such as enhancing roadway right-of-ways, incorporating small pocket parks into urban plans, and supporting greater densities of street trees. For example, the insect conservation organization Xerces has developed best management practices for transportation agencies to support pollinators within highway rights-of-way. These best management practices, such as reducing mowing frequency, planting native species, and reducing herbicide use in rights-of-way, can help pollinators move through the landscape for daily foraging or for dispersal between larger habitat patches. In addition, strategies such as "*pocket parks*" and other small greening initiatives have been shown to enhance bird species richness in urban areas (Strohbach *et al.* 2013), especially if these strategies are part of broader greening initiatives. Other creative city interventions, such as Sheffield, England's "Grey to Green Corridor," New York City's "High Line," and Chicago's "Bloomingdale Trail," convert abandoned spaces such as rail lines or redundant roadways into corridors planted with native vegetation that can make the dense urban matrix more hospitable and traversable to mobile species such as birds and bees.

Several cities have taken creative approaches to enhancing green infrastructure networks on private lands, such as the "Pollinator Pathway" project in Seattle, Washington. The Pollinator Pathway pilot project is a 1.6 km long, 3.6 m wide corridor of pollinator-friendly plantings in front of residential yards. The pathway creates a backbone of connectivity between two green spaces near downtown Seattle. It is especially beneficial when these kinds of approaches incorporate native plants species that are also found in nearby habitat remnants. This helps to better "fit" the small-scale green infrastructure within the local ecosystem and broader habitat network (Cerra and Crain 2016).

It is important to note that green infrastructure enhancements within one land use type can generate effects that *spill over* or extend into other adjacent or nearby areas. For example, increasing floral resources in home gardens could increase pollination services at community gardens (Davis *et al.* 2017). Enhancing habitat in home gardens, boulevards, and utility rights-of-way could extend the width of riparian corridors and functionally connect large green spaces (Rudd *et al.* 2002). And conversely, connecting home gardens by woody corridors could enhance biodiversity in the home gardens themselves (Vergnes *et al.* 2012).

TABLE 12.1 Different kinds of green space in the urban green infrastructure network. Size and spatial coverage data are specific to Chicago, Illinois (USA).

Green space	Approx. size of individual green space	Approx. spatial coverage	Contributions to biodiversity conservation	Typical management activities	Possible ecological problems and disservices
Home gardens	~50 m²	3,000 ha[†]	High plant diversity and density of floral resources; may be especially good habitat for insects and birds	Can include herbicides, pesticides, fertilizers, mowing. Might have high temporal turnover of plant species (and soil disturbance) due to gardening activities	Perceived garden or house pests (e.g., rodents, deer), or 'scary' animals (wasps, spiders, etc.), non-native plant species
Cemeteries	<1 – 131 ha	700 ha	Larger habitats with reduced human disturbance due to walls and restricted hours; diversity of substrates for lichen; older, larger trees than typically found in cities	Relatively frequent and intense management: mowing, pruning, chemical fertilizers and/or pesticides	Chemical inputs, non-native plant species, headstones as ecological traps
Vacant lots	~300 m², but highly variable	780 ha	Relatively low-disturbance habitat, potential for pollinator provisioning and stopover habitat	Relatively little management: occasional mowing and/or herbicide	Harboring unwanted pests and/or weed species
Urban agriculture sites	<5 – 24,000 m²	26 ha	High plant diversity and structural heterogeneity; rooftop and container gardens can provide green infrastructure in highly developed areas	Herbicide and/or insecticide use (although typically less than in residential gardens); hand weeding; chemical fertilizers and/or manure	Insect pests are often actively removed, but gardens may provide easy food source for vertebrate pests such as rats
City parks	<1 – 500 ha	3200 ha	Often larger habitats, some include native plant gardens	Can range from ecological restoration to frequent and intense management: mowing, pruning, chemical fertilizers and/or pesticides	"Lawn" type parks and sports fields may not provide useful habitat for many species. Some species might be isolated in natural areas surrounded by high human recreation or traffic.

[†]rough estimate based on ~30,000 ha of residential land in Chicago

There is clearly a benefit to integrating the design and management of home gardens, urban agriculture, vacant lots, and cemeteries into city-wide biodiversity strategies. However, this task is complicated by the fact that different stakeholders have different goals and constraints. Many urban green spaces are managed at the local scale, but collaboration and communication between a range of stakeholders across all sectors of society is required to maximize the ecological potential of green infrastructure. The concept of "*scale-crossing brokers*" (Ernstson *et al.* 2010), in which individuals or organizations help make connections between local- and landscape-scale processes, can help address the mismatch between stakeholders operating at different scales and thereby maximize conservation of urban biodiversity.

References

Anderson E. and Minor E.S. (2017) Vacant lots: An underexplored resource for ecological and social benefits in cities. *Urban Forestry & Urban Greening* 21: 146–152.

Barrett G.W. and Barrett T.L. (2001) Cemeteries as repositories of natural and cultural diversity. *Conservation Biology* 15: 1820–1824.

Basmajian C. and Coutts C. (2010) Planning for the disposal of the dead. *Journal of the American Planning Association* 76: 305–317.

Belaire J.A., Whelan C.J. and Minor E.S. (2014) Having our yards and sharing them too: The collective effects of yards on native bird species in an urban landscape. *Ecological Applications* 24: 2132–2143.

Benedict M. and McMahon E.T. (2006) *Green Infrastructure: Linking Landscapes and Communities*. Island Press, Washington, DC.

Bergey E.A. and Figueroa L.L. (2016) Residential yards as designer ecosystems: Effects of yard management on land snail species composition. *Ecological Applications* 26: 2538–2547.

Bernholt H., Kehlenbeck K., Gebauer J. and Buerkert A. (2009) Plant species richness and diversity in urban and peri-urban gardens of Niamey, Niger. *Agroforestry Systems* 77: 159–179.

Brown K.H. and Jameton A.L. (2000) Public health implications of urban agriculture. *Journal of Public Health Policy* 21: 20.

Burghardt K.T. and Tallamy D.W. (2013) Plant origin asymmetrically impacts feeding guilds and life stages driving community structure of herbivorous arthropods. *Diversity and Distributions* 19: 1553–1565.

Cerra J. and Crain R. (2016) Urban birds and planting design: Strategies for incorporating ecological goals into residential landscapes. *Urban Ecosystems*: 1–24.

Cockle K.L., Martin K. and Wesolowski T. (2011) Woodpeckers, decay, and the future of cavity-nesting vertebrate communities worldwide. *Frontiers in Ecology and the Environment* 9: 377–382.

Colding J. (2007) "Ecological land-use complementation" for building resilience in urban ecosystems. *Landscape and Urban Planning* 81: 46–55.

Curran W. and Hamilton T. (2012) Just green enough: Contesting environmental gentrification in Greenpoint, Brooklyn. *Local Environment* 17: 1027–1042.

Daily G.C. (1997) *Nature's Services: Societal Dependence on Natural Ecosystems*. Island Press, Washington, DC.

Davies Z.G., Fuller R.A., Loram A., Irvine K.N., Sims V. and Gaston K.J. (2009) A national scale inventory of resource provision for biodiversity within domestic gardens. *Biological Conservation* 142: 761–771.

Davis A.Y., Lonsdorf E.V., Shierk C., Matteson K.C., Taylor J., Lovell S.T. and Minor E.S. (2017) Enhancing pollination supply in an urban ecosystem through landscape modifications. *Landscape and Urban Planning* 162: 157–166.

Dooling S. (2009) Ecological gentrification: A research agenda exploring justice in the city. *International Journal of Urban and Regional Research* 33: 621–639.

Ernstson H., Barthel S., Andersson E. and Borgström S.T. (2010) Scale-crossing brokers and network governance of urban ecosystem services: the case of Stockholm. *Ecology and Society* 15(4): 28.

European Commission (2013) Communication from the Commission to the European Parliament, The Council, The European Economic and Social Committee and the Committee of the Regions: Green Infrastructure (GI)—Enhancing Europe's Natural Capital. Document 52013DC0249.

Fischer J. and Lindenmayer D.B. (2007) Landscape modification and habitat fragmentation: a synthesis. *Global Ecology and Biogeography* 16: 265–280.

French S. (1974) The cemetery as cultural institution: The establishment of Mount Auburn and the "rural cemetery" movement. *American Quarterly* 26: 37–59.

Gallagher J. (2010) *Reimagining Detroit: Opportunities for Redefining an American City.* Wayne State University Press, Detroit, MI.

Gardiner M.M., Burkman C.E. and Prajzner S.P. (2013) The value of urban vacant land to support arthropod biodiversity and ecosystem services. *Environmental Entomology* 42(6): 1123–1136.

Goddard M.A., Dougill A.J. and Benton T.J. (2009) Why garden for wildlife? Social and ecological drivers, motivations and barriers for biodiversity management in residential landscapes. *Ecological Economics* 86: 258–273.

Grimm N.B., Faeth S.H., Golubiewski N.E., Redman C.L., Wu J., Bai X. and Briggs J.M. (2008) Global change and the ecology of cities. *Science* 319: 756–760.

Horváth G., Malik P., Kriska G. and Wildermuth H. (2007) Ecological traps for dragonflies in a cemetery: the attraction of Sympetrum species (Odonata: Libellulidae) by horizontally polarizing black gravestones. *Freshwater Biology* 52: 1700–1709.

Kendal D., Williams K.J.H. and Williams N.S.G. (2012) Plant traits link people's plant preferences to the composition of their gardens. *Landscape and Urban Planning* 105: 34–42.

Kremer P. and DeLiberty T.L. (2011) Local food practices and growing potential: Mapping the case of Philadelphia. *Applied Geography* 31: 1252–1261.

Kremer P., Hamstead Z.A. and McPherson T. (2013) A social–ecological assessment of vacant lots in New York City. *Landscape and Urban Planning* 120: 218–233.

Larson J.L., Kesheimer A.J. and Potter D.A. (2014) Pollinator assemblages on dandelions and white clover in urban and suburban lawns. *Journal of Insect Conservation* 18: 863–873.

Lerman S.B., Turner V.K. and Bang C. (2012) Homeowner associations as a vehicle for promoting native urban biodiversity. *Ecology and Society* 17: 45.

Lin B.B., Philpott S.M. and Jha S. (2015) The future of urban agriculture and biodiversity–ecosystem services: Challenges and next steps. *Basic and Applied Ecology* 16: 189–201.

Loram A., Thompson K., Warren P. and Gaston K.J. (2008) Urban domestic gardens (XII): The richness and composition of the flora in five UK cities. *Journal of Vegetation Science* 19: 321–330.

Lowenstein D.M., Matteson K.C., Xiao I., Silva A.M. and Minor E.S. (2014) Humans, bees, and pollination services in the city: The case of Chicago, IL (USA). *Biodiversity and Conservation* 23 (11): 2857–2874.

Lowenstein D.M. and Minor E.S. (in review) Influence of top-down and bottom-up forces on herbivorous insects in urban agriculture. *Biological Control.*

Lussenhop J. (1977) Urban cemeteries as bird refuges. *The Condor* 79: 456–461.

Magle S.B., Reyes P., Zhu J. and Crooks K.R. (2010) Extirpation, colonization, and habitat dynamics of a keystone species along an urban gradient. *Biological Conservation* 143: 2146–2155.

Matteson K.C., Ascher J.S. and Langellotto G.A. (2008) Bee richness and abundance in New York City urban gardens. *Annals of the Entomological Society of America* 101: 140–150.

Pattison W.D. (1955) The cemeteries of Chicago: A phase of land utilization. *Annals of the Association of American Geographers* 45: 245–257.

Pediaditi K., Doick K.J. and Moffat A.J. (2010) Monitoring and evaluation practice for brownfield, regeneration to greenspace initiations: A meta-evaluation of assessment and monitoring tools. *Landscape and Urban Planning* 97(1): 22–36.

Renton M., Shackelford N. and Standish R.J. (2014) How will climate variability interact with long-term climate change to affect the persistence of plant species in fragmented landscapes? *Environmental Conservation* 41(2): 110–121.

Rudd H., Vala J. and Schaefer V. (2002) Importance of backyard habitat in a comprehensive biodiversity conservation strategy: A connectivity analysis of urban green spaces. *Restoration Ecology* 10: 368–375.

Robbins P., Polderman A. and Birkenholtz T. (2001) Lawns and toxins: An ecology of the city. *Cities* 18(6): 369–380.

Shelton T.V. (2008) Unmaking historic spaces: Urban progress and the San Francisco cemetery debate, 1895–1937. *California History* 85(3): 26–70.

Smith R.M., Warren P.W., Thompson K. and Gaston K.J. (2006) Urban domestic gardens (VI): Environmental correlates of invertebrate species richness. *Biodiversity & Conservation* 15: 2415–2438.

Strohbach M.W., Lerman S.B. and Warren P. (2013) Are small greening areas enhancing bird diversity? Insights from community-driven greening projects in Boston. *Landscape and Urban Planning* 114: 69–79.

Taylor J.R. and Lovell S.T. (2012) Mapping public and private spaces of urban agriculture in Chicago through the analysis of high-resolution aerial images in Google Earth. *Landscape and Urban Planning* 108: 57–70.

Thebo A.L., Drechsel P. and Lambin E.F. (2014) Global assessment of urban and peri-urban agriculture: Irrigated and rainfed croplands. *Environmental Research Letters* 9: 114002.

Threlfall C.G., Ossola A., Hahs A.K., Williams N.S.G., Wilson L. and Livesley S.J. (2016) Variation in vegetation structure and composition across urban green space types. *Frontiers in Ecology and Evolution* 4: 66.

Tischendorf L. and Fahrig L. (2000) On the usage and measurement of landscape ecology. *Oikos* 90: 7–19.

Tornaghi C. (2014) Critical geography of urban agriculture. *Progress in Human Geography* 38: 551–567.

van Heezik Y., Freeman C., Porter A. and Dickinson K.J.M. (2013) Garden size, householder knowledge, and socio-economic status influence plant and bird diversity at the scale of individual gardens. *Ecosystems* 16: 1442–1454.

Vergnes A., Le Viol I. and Clergeau P. (2012) Green corridors in urban landscapes affect the arthropod communities of domestic gardens. *Biological Conservation* 145: 171–178.

Vessel M.F. and Wong H.H. (1987) *Natural History of Vacant Lots.* University of California Press, Berkeley, CA.

Wachholder B., Burmeister M.S., Methven A.S. and Meiners S.J. (2004) Biotic and abiotic effects on lichen community structure in an Illinois cemetery. *Erigenia: Journal of the Southern Illinois Native Plant Society* 20: 29–36.

Widows S.A. and Drake D. (2014) Evaluating the National Wildlife Federation's Certified Wildlife Habitat program. *Landscape and Urban Planning* 129: 32–43.

13

DESIGNING NATURE IN CITIES TO SAFEGUARD MEANINGFUL EXPERIENCES OF BIODIVERSITY IN AN URBANIZING WORLD

Assaf Shwartz

Introduction

Human appropriation of natural resources has led to an unprecedented biodiversity crisis that conservation efforts are not managing to contain. Since human activity is driving this crisis, the solutions to it will largely depend on our actions in the future. Protecting pristine wilderness in nature reserves has long been used to mitigate these threats by separating biodiversity from incompatible land uses. But current evidence indicates that considerable increase in the global coverage of protected areas is not sufficient to halt the biodiversity crisis (Mora and Sale 2011). If we insist on putting our endeavors only on conserving things that look like pristine wilderness, our best efforts may only delay their destruction (Marris 2013). Today, there is a growing understanding that conservation efforts need to be expanded to the vast majority of land that is directly influenced by human activity (83% of the total land surface). This means seeking for solutions that reduce our detrimental impacts on the environment and that enable coexistence between humans and other species.

Urban development is possibly the greatest example of human-induced environmental degradation, leading to some of the highest species extinction rates while isolating the majority of the world population from nature. Although agriculture is certainly the most spatially extensive form of land-use alteration, urbanization is one of the most destructive forms of global environmental change that is currently growing faster than any other type of land-use (Lin and Fuller 2013). While urban living has many major societal and economic benefits, it also means that more people live in biologically impoverished environments, far from nature and spend most of their lives indoors. This reduced experience of nature in people's daily lives is an important environmental issue, because it can transform the way people value the natural world around them, seriously endangering the potential to stimulate the behavioral change and public support needed to achieve conservation objectives (Miller 2005; Dunn *et al.* 2006). It is also profoundly concerning given the mounting empirical evidence showing that interaction with nature delivers measurable physical, psychological and social benefits for people (see Chapter 8). Designing biodiverse biophilic cities that protect and restore biodiversity, while expanding the opportunities to interact with it, was suggested as a means to mitigate the various detrimental impacts of urbanization (Beatley 2011; Shwartz *et al.* 2014a). Building on this framework (Figure 13.1), this chapter will first explore the causes and consequences of human alienation from nature. It will then present and discuss how different biophilic urban designs, management strategies, methods and tools can facilitate traditional and novel interactions between people and biodiversity, to achieve better coexistence between people and nature.

The extinction of experience causes and consequences

Since the dawn of human civilization, the profound connection of people to nature has been something quite clear. Yet today, we are more and more entrenched in the view that we have been able to overcome the need for nature (Beatley 2011). As a result, we have degraded and destroyed a large part of the nature surrounding us and also needlessly excluded it from our daily lives. Although the evidence on the importance of integrating nature into our lives is mounting, we still lack the broad-based public support to encourage local and central governments to take action and integrate nature conservation considerations into policies. Interestingly, the same processes that threaten biodiversity, such as urbanization, agricultural intensification

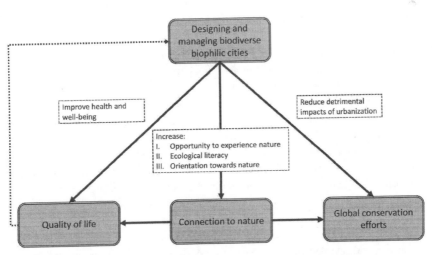

FIGURE 13.1   Conceptual framework identifying the potential benefits to city-dwellers and biodiversity conservation of designing and managing biodiverse biophilic cities. The dotted lines represent feedback from outcomes to design principles. Dashes squares provide explanations of the underlying mechanisms.

and biotic homogenization, also increasingly isolate humans from the experience of the natural world surrounding them. Robert Pyle, who coined the phrase *"extinction of experience"* in 1978 to describe this alienation from nature, argued that this process is one of the greatest causes of the biodiversity crisis, as collective ignorance leads to collective indifference (Pyle 1978, 2003). Almost four decades have passed since Pyle first suggested his theory and despite greater awareness, information is still scarce about the causes, consequences and most importantly the potential solutions that could help mitigating the extinction of experience (Soga and Gaston 2016).

The loss of opportunity to directly experience nature and the loss of positive orientation towards engaging with nature (reduced emotional affinity with nature) have been suggested as the main causes of the extinction of experience (Miller 2005; Soga and Gaston 2016). Today, the majority of the world's population lives in biologically impoverished cities and people spend most of their time indoors, with limited opportunity to interact with nature in their day-to-day life (Lacoeuilhe *et al.* 2017). Urban densification aggravates this problem by increasing the geographic distances from natural environments and green spaces, resulting in lower visiting frequencies to those spaces (Soga *et al.* 2015). Reduced opportunity may lead to a diminution in ecological knowledge and literacy, the loss of people's orientation towards engaging with nature and their ability to develop emotional attachment with nature. This lower orientation and affinity with nature may also influence people's motivation to experience it, as reduced emotional connectedness to nature was also found to decrease visiting frequency and time spent in natural environment (Soga and Gaston 2016). This vicious cycle can be moderated by several variables and particularly childhood experience of nature, because lack of emotional attachments during childhood, is typically carried over into adulthood.

In his book *Last Child in the Woods,* Richard Louv (2008) has argued that humans and especially children, are suffering from a *"nature deficit disorder,"* resulting in a wide range of behavioral problems. The lives of children today are much more programmed and scheduled than just a generation ago. They spend much time inside, in front of televisions and computers, with little freedom to explore nature also due to increasing parental concerns about their safety (the *"Bogeyman Syndrome Redux"*; Louv 2008). All the above factors are contributing to the increasing disconnection of children from nature. Soga and Gaston (2016) have recently summarized some temporal evidence that demonstrates how interaction with nature has changed among children and adults over a decade in several place across the globe. For instance, in the US, the share of children participating in outdoor activity decreased by 6 percent, from 16 percent in 1997 to 10 percent in 2003 and significant reduction in the amount of time spent outside was also recorded (from 36 to 25 minutes per week accordingly). Even indirect experience of nature during indoor activities (e.g. watching a movie) is diminishing, as revealed by an analysis of outdoor environments representation over 70 years of Disney animated films (Couvet and Prevot 2015). When these experiences are reduced, the ability to generate emotional attachments is not carried over into adulthood, resulting in some substantial impacts on both people and conservation.

The extinction of experience is recognized as a major contemporary social and environmental issue (Miller 2005; Soga and Gaston 2016). Numerous studies have demonstrated the wide range of health and well-being benefits associated with the exposure to or interaction with nature (see also Chapter 8). Loss of these interactions will thus result in reductions in these associated benefits. For instance, van den Berg and colleagues (2010) have demonstrated that participation in community gardening in several large cities in the Netherlands was related to improved physical and psychological conditions. Even a glimpse of nature viewed in photographic images and videos of a natural landscape have been shown to reduce skin conductance, heart rate and other physiological indicators of stress (Ulrich *et al.* 1991; Gladwell *et al.* 2012). Although most studies examine short-term health benefits, few studies have also documented the long-lasting influence of exposure to nature on health for instance on diabetes, heart diseases and more generally the longevity of senior citizens (reviewed by Soga and Gaston 2016). Exposure to nature can also have significant influence on cognitive abilities and recently Bratman and colleagues (2015) have demonstrated how a 50 minute walk in nature can improve memory and executive attention. Notwithstanding the important evidence these studies offer on the role of nature in providing multiple social benefits, their value for conservation remains limited, since they study the experience of nature as a "*black box.*" In other words, the aspects of nature that convey these benefits are not yet understood.

An overview of studies exploring the relationship between biodiversity (as a measure of nature's complexity) and well-being unveiled the "*people-biodiversity paradox.*" This paradox comprises a fundamental mismatch between: (i) the preferences people say they have for biodiversity and how this relates to their self-reported well-being; and, (ii) people's limited ability to perceive biodiversity, which may prevent them from benefiting fully from direct interaction with nature's complexity (Pett *et al.* 2016). The Biophilia Theory, which proposes an innate human tendency to affiliate with nature (Wilson 1984), can explain the positive affinity of people towards biodiversity, while the lack of ecological literacy and knowledge can be attributed to the extinction of experience (Shwartz *et al.* 2014b). This tendency is not a "*hard wired*" instinct, but rather a genetically programed inclination that depends on adequate learning and experiences to functionally occur. The extinction of experience can thus create a cycle of impoverishment, in which increased estrangement between people and nature may affect people's capability to obtain the "*dose of nature*" needed to ensure their health well-being (Shanahan *et al.* 2016), but also change their emotions, attitudes and behaviors toward nature.

Beyond the social consequences of the extinction of experience, the lack of knowledge of biodiversity is a major concern, because "*people who care conserve; people who don't know don't care*" (Pyle 2003). A recent study has demonstrated that people's positive or negative attitudes towards garden birds were related to their knowledge and familiarity with the species (Cox and Gaston 2015). The theory of planned behavior can help formulate relationships between knowledge, attitudes and behaviors (Ajzen 1991). Knowledge is often presented as a key step for people to adopt pro-environmental behaviors, among other factors such as practicality, ethics and emotions. The lack of interactions with nature during childhood and adulthood, can reduce people's ecological knowledge and thus lead to changes in emotions, attitudes and behaviors toward nature. Thus, the likelihood that one would support conservation initiatives and engage in conservation behaviors as an adult can be predicted by the amount of time and the quality of nature experience they have had, allowing them to develop a stronger sense of place. A key issue that exacerbates this problem is that there are several likely feedback cycles causing alienation and indifference toward nature via loss of opportunity and orientation (Soga and Gaston 2016). For instance, the biologically impoverished environment encountered today by many urban children becomes the baseline against which future socio-ecological degradation is assessed (Miller 2005). This "*shifting baseline syndrome*" (Papworth *et al.* 2009) leads to reduced expectations, as people do not recognize what has been lost.

A question that arises is how we can break this "*cycle of impoverishment*" (Miller 2005). Conserving or enhancing biodiversity in the places where people live and work can be an important first step. However, such efforts to increase the opportunity to encounter nature may not be enough to avert the extinction of experience, because people might not have the interest or ability to notice the rich nature available in those environments (Dallimer *et al.* 2012; Shwartz *et al.* 2014b). There is thus a need for solutions that will make nature and its complexity more accessible for people, and foster a strong emotional relation. The following sections critically explore the outcomes of different urban designs, opportunities, activities and strategies that can either establish, or enhance, both traditional and novel meaningful experiences of nature in cities.

Planning and managing cities to avert the extinction of experience

Although nature has always been an integral part of cities, it has usually been shaped for people, reflecting their cultural values and with little consideration for conservation. Urban green spaces have a long history in the development of different civilizations across the globe. The origin of gardens was first driven by religious beliefs, in an attempt to create some paradise on Earth (e.g. Nebamun's garden and Persian's walled gardens). In the eighteenth century, the French architect Ledoux described a city "whose neighborhoods, dedicated to peace and happiness, would be planned with gardens rivalling Eden" (Ledoux 1804, p. 1). City planners have long ago grasped the importance of these green spaces to improve the quality of life of city dwellers. Consequently, they integrated manicured nature in the form of parks and gardens in the urban fabric, designed to reflect aesthetic values and to provide recreation. Other types of green spaces (e.g. remnants, wastelands) are also being created intentionally or unintentionally, as a consequence of socio–economic and political dynamics. More recently, under the accelerating urbanization, green areas are becoming increasingly valued by city-dwellers. This has been leading to the development of new greener urban environments in suburban or exurban areas that are characterized by a relatively high percentage of private gardens. But even in more dense urban areas, practitioners are increasingly seeking for the means to create "green" grey infrastructures, notably on buildings with vertical plant walls and green roofs, but also in streetscapes by planning pervious parking lots for instance.

Altogether, these green infrastructures cover a non-negligible part of our cities and they also have the potential to host a rich biodiversity (Fuller and Gaston

2009). The grandest challenge today is to discover how to plan and manage these green (and sometime grey) infrastructures sustainably to accommodate the "dose" of nature needed to: (i) minimize the detrimental ecological impacts of urbanization; (ii) ensure people's health and well-being directly and indirectly through the provision of ecosystem services and (iii) integrate this nature into our daily lives to propel adults and children to experience and cherish it, averting the extinction of experience. Cities have thus the potential to serve as "urban arks – places that help to counter-balance the diminished (and diminishing) biodiversity outside of cities" (Beatley 2011). But the creation of these *urban arks* may appear to be conflicting with more regional biodiversity objectives, since integrating complex and rich nature in cities is land demanding, and may occur at the expense of nearby greener environments. Therefore, considering urban development forms and their consequences on biodiversity and its experience is key for urban and regional planning.

Urban form, biodiversity and its experience

Cities have become the main habitat for humans. Today the net population growth is occurring almost entirely within urban areas that will need to accommodate an additional 2.6 billion people by 2050 (Lin and Fuller 2013). As such, a question arises: how can we modify cities to reduce their impacts on biodiversity and enhance the experience of nature and its benefits. In search of an answer, Lin and Fuller (2013) suggested applying to urban areas the land sparing/land sharing paradigm, previously developed to explore how to balance food production and biodiversity conservation. Accordingly, urban forms can be placed on a continuum ranging from: (i) sparing or compacting urban development that reduces the spatial extent of developed areas by intensifying urbanization and population density (Figure 13.2). In this case, ecological impact is heightened locally, but remains spatially constrained at the landscape or regional scale. Sharing, green or sprawl development that minimizes the intensity of urbanization, so that ecological impacts are reduced locally (Figure 13.2), but in turn urban development must be spread over large areas to accommodate the increasing urban population growth. To date, few studies have empirically explored the ecological benefits of sparing vs. sharing in urban areas. These studies have demonstrated that, if well planned, land sparing (compact development) performed better in conserving regional biodiversity than land sharing (e.g. Sushinsky *et al.* 2013; Soga *et al.* 2014). This is because compact urban form profoundly reduces the amount of land consumed for dwelling and thus enables the establishment of large green spaces or remnant native habitats that suits many urbanophobe species that normally avoid cities.

These types of urban forms also have consequences on people's access to, and experience of nature. Under land sharing, neighborhoods' greenness is high and green spaces tend to be smaller and more fragmented. Biotic communities in those habitats, such as private yards and public parks, tend to be less biodiverse than in large green spaces (Shwartz *et al.* 2013). However, under this urban development form, biodiversity is more evenly spread across the city and on average closer to residents. In contrast, under land sparing, a rich biodiversity is concentrated into a

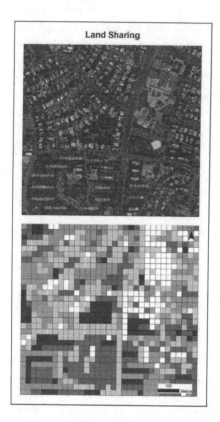 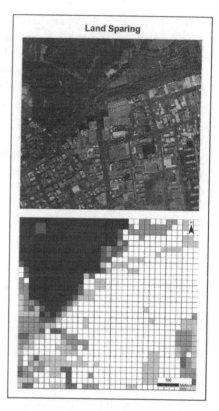

FIGURE 13.2 Schematic illustration of land-sharing or sprawl and land-sparing or compact development forms in the Tel-Aviv metropolis, Israel (25 m² cell size). Darker pixel shading represents with higher vegetation cover. (*Source: Assaf Shwartz, unpublished.*)

few, large, green spaces and this increases the geographic distances to natural environments within and outside cities. As a result, under a land sparing development form, people's accessibility to green spaces is lower than under land sharing and this can lead to a reduction in visiting frequency and duration of visits to green spaces, as was found in Tokyo, Japan (Soga *et al.* 2015). As such, land sparing can profoundly reduce the experience of nature in people's daily lives resulting in associated negative consequences on health, well-being and affinity towards nature and conservation.

These insights seem to lead to an inevitable conflict between regional biodiversity objectives and local, social and conservation considerations. Yet, cities are built for humans and given the growing recognition of the importance of interaction with nature on people's lives, at least in western societies, green infrastructures are and will be integrated in any type of urban development form (Figure 13.3a). The question that remains is how these green infrastructures (mentioned above) will be planned and managed? If our cities will be designed with nature in mind, to cherish what already exists, restore what is possible and insert biodiversity into the streets,

FIGURE 13.3 Examples of: (a) compact urban development form (Singapore) that can free area to create large multifunctional green spaces. The way these green spaces are planned and managed determines their quality for conserving biodiversity and allowing meaningful interaction with nature; (b) tiny road signs placed in Vingis Park (Vilnius Lithuania) to draw people's attention to unseen or unnoticed wildlife; (c) insect hotel—a man-made structure created to provide shelter and nesting resources for insects and (d) bat billboard—an interactive billboard that create artificial habitats for bats, but also facilitate interactions between bats and humans.

building, parks, private and public gardens (Beatley 2011), then even dense cities will be able to avert the extinction of experience. This will align the agendas of public health and conservation (Pett *et al.* 2016). A key step in achieving this goal is to enhance the natural capital in the places where most people live and work today, so as to provide opportunities for interacting with nature and its complexity.

Unexploited opportunities

Regardless of their form, cities and neighborhoods can be planned and designed to better fit around and conserve natural elements (Beatley 2011). All types of open and built spaces between buildings and streets, from large remnants and parks to small roundabouts and private gardens, represent many promises to enhance biodiversity in cities. This is especially true in compact urban developments, in which it is possible to free some land for the benefit of inserting nature into cities (Figure 13.3a). There is a growing consensus among planners and scholars that compactness, mixed land uses and diversity of activity has an important role in achieving sustainable cities (Jabareen 2006). Designing multifunctional green spaces or a structured array of mixed land uses can help turning cities into novel ecosystems that are ecologically sound, aesthetically satisfying, economically rewarding and favorable to the continued growth of civilization (Marris 2013). Although there is rarely a solution that is universal, knowledge about the ecosystem that dominated the area prior to the urban development is often an important first step.

Completely restoring the original ecosystems is most probably impossible, but knowledge about these can inspire practitioners and can help them design and manage those spaces to achieve relevant conservation goals (Marris 2013). Such designs can also provide many other ecosystem services, such as reducing pollution, urban heat, storm water runoff and, if access to this nature is maintained, several other societal benefits. However, one size does not fit all, and in some cases, different ecosystem services can conflict between each other and with other conservation or social objectives (Shwartz *et al.* 2014a). For instance, lawns are one of city dwellers' favorite types of land covers in urban green spaces. Yet, well-maintained lawns, which provide recreation services, offer a poor habitat for many species (i.e. poor nature conservation services). An evident solution can be to reduce their maintenance or even rewild the lawns into native prairies, but such interventions may not be appreciated, as these novel lawns may look unaesthetic or messy (see Chapter 14). Using "*cues of care*" in design, such as intensively mowing a strip along the paths or edges of the novel lawn, can enhance acceptance by placing the ecological functions in a recognizable cultural context (Nassauer 1995). Nevertheless, such interventions may not be sufficient to modify people's acceptance, if the induced changes significantly alter the way people make use of the lawn (e.g. sports, sunbathing).

Understanding the ecological setting is one first important step, but different designs and management interventions should be carefully considered against ecological and social goals to find the middle ground that allows establishing sustainable, multifunctional, urban ecosystems. To find this optimal balance, which is often case specific, urban planners and green space managers can consider different interventions that locally spare or share land uses, either in time or space. The management of lawns in Parisian gardens exemplifies a plethora of such possible solutions. In some gardens, geophytes of *Narcissus sp.* and other species are planted within lawns. During the winter and early spring the entrance to those lawns is restricted, but people can benefit from the colorful blossom of several species. Later in the spring and summer, these lawns are mowed and opened to the public, so that people can benefit from using them intensively. Alternatively, in some gardens, a portion of the lawns has been converted to flowering prairie, while the remaining parts are managed extensively (i.e. with no pesticides, low mowing frequency and high mowing height). Altogether, these practices facilitate the conditions for wild plants and pollinator species and therefore host rich diversities of these and other taxonomic groups (Shwartz *et al.* 2013). Recognizing that wild animals and plants are also legitimate stakeholders in the design and management of cities is an important step in creating biophilic cities that cherish nature and foster interaction with it.

A significant amount of urban conservation research has been directed to understand how different local- and landscape-scale design variables influence biodiversity in the urban environment (Goddard *et al.* 2010). At a local scale, there is a consensus among urban ecologists that biodiversity increases with patch area. But land is a limited and expensive commodity in urban areas, and as it was demonstrated above, expanding cities to establish large green spaces may not be ecologically justifiable. There is therefore a need for creative thinking about how to create spaces that are big enough to host viable nature. Greening grey infrastructures such as roofs, walls and streets can help establish a network of multifunctional spaces that together can form areas that are large enough to support rich biodiversity and facilitate people's interaction with it. For instance, in Paris and New York, such parks were established in unusual locations on the top of obsolete elevated railway lines (Promenade Plantée and the High Line). In Paris, this viaduct is about 4.5 km long and it connects several green spaces to the largest green park in Paris, the Bois de Vincennes, allowing the passage and interaction of wildlife and people. A recent meta-analysis of 75 cities worldwide covering a large variety of taxonomic groups has demonstrated that together with patch area, such corridors have the strongest positive effect on biodiversity (Beninde *et al.* 2015).

Encouraging plant- and wildlife-friendly management of existing green spaces and left over-spaces in neighborhoods and even in industrial areas can insert significant nature into our cities and our lives with minimal spatial cost (Lacoeuilhe *et al.* 2017). Identifying and minimizing different forms of disturbances, such as light and noise pollution, is an important first step that is often overlooked in efforts to improve the conditions for many species in cities. In many cases, such interventions can also co-benefit people directly (e.g. more peaceful environment, reduced municipal taxes) and indirectly by improving the environmental conditions in the city (e.g. reduce CO_2 emissions). Providing shelter, nesting and feeding resources is also essential in maintaining or even enhancing nature in cities. Mounting evidence has demonstrated that this can be achieved by careful consideration of the type of vegetation used, its configuration (e.g. vegetation complexity and tree cover, habitat diversity) and the intensity of management (e.g. pesticide use and mowing frequency) (Goddard *et al.* 2010; Beninde *et al.* 2015).

One good example of such initiative is the *differential management* program. First developed in Germany during the 1990s, it represented an alternative to intensive horticultural management of urban green spaces and it is now widely spread across central cities in Europe (e.g. Amsterdam, Hamburg, Brussels and Paris). The program promotes a range of design and management interventions for developing sustainable green spaces in urban areas that have been demonstrated to be successful in increasing biodiversity, even in small gardens (Shwartz *et al.* 2013). Finally, almost anywhere in cities there is an option to supplement food and nesting resources artificially to a variety of species (Figure 13.3b). Although such interventions can serve as an excellent means to increase the accessibility of nature, they should be considered carefully to avoid the creation of ecological traps, leading species to settle in poor quality habitats that cannot sustain their populations viably. Sustainable novel urban ecosystems should therefore be designed and managed not only to host a rich biodiversity, but also to make it more viable and visible (Miller 2005).

Although cities are characterized by highly modified ecosystems, they still contain a variety of unexploited opportunities that, with suitable creative planning and management, can turn into urban arks—places we co-inhabit with myriads of other life forms (Beatley 2011). But public acceptance and support of such initiatives is not yet guaranteed, because people already demonstrate reduced orientation towards nature and poor ecological skills to experience and even benefit from its complexity (Pett *et al.* 2016; Soga and Gaston 2016). Integrating more biodiversity into cities can only take us part of the way. If we are to change the direction of the extinction of experience cycle from impoverishment to enrichment, we must also invest in creating meaningful experiences of nature, and acknowledge and cope with negative interactions.

Designing meaningful experiences of nature

Addressing social and cultural obstacles in designing and managing biophilic cities may be significantly harder than resolving ecological challenges. Even when excellent opportunities to experience nature exist in urban areas, there is no guarantee that residents, children and adults alike, will actually use and benefit from them. There are multiple reasons for this (Beatley 2011). Today in societies that follow a Western way of life, most people do not feel the need to interact with nature for biological reasons anymore. Humans spend most of their time indoors, with busy schedules, heavily programmed lives dominated by screen time (televisions, computers and mobile phones). When people do get out, they often consciously experience only a small fraction of the natural world surrounding them, given their limited ecological abilities (Dallimer *et al.* 2012; Shwartz *et al.* 2014b), as it was reflected by the words of a participant in an interview about nature in a Parisian public garden: "we go to the park, we do almost the same thing every time, and it's true, we don't necessarily realize all these things going on around us" (Shwartz *et al.* 2012). Fear of some elements of nature, and especially parental concerns, is another factor that significantly impedes our contact with the natural world, when adults and children are already outdoors (Louv 2008). Therefore, the challenge goes beyond the creation of cities that host rich and complex nature. We should seek for solutions that make this biodiversity accessible and not threatening when possible, but safe and harmless when needed, to create a set of novel experiences of nature (Clayton *et al.* 2017).

Many of the interventions mentioned above can provide the basic opportunities for meaningful nature experiences to occur. But compared to mere observation, interactive, multisensory human–nature experiences that engage emotions and create more lasting memories are needed to avert the extinction of experience (Clayton *et al.* 2017). Such interactions are often shaped and influenced by the landscapes in which they take place, by products and services mediating these experiences, and by the common beliefs and stories we tell ourselves about the nature with which we come into contact (Metcalfe 2015). A lot happens at the meeting point between humans and biodiversity. Empathy, affection and wonder can be sparked; conflicts may arise; there is a process of mutual shaping and reshaping, there is caring and nursing, trampling and cruelty, fear and respect. Biodiversity can be enhanced in small and large, public and private, green and grey infrastructures in compact or sprawled cities. But when biodiversity encounters become more frequent in urban areas, they can either be shaped by design and education, or left to chance. Crafting these encounters has the potential of reducing conflicts, shifting perceptions and attitudes towards biodiversity, and preparing the ground for more inclusive and biodiverse novel ecosystems (Metcalfe 2015).

Much of urban biodiversity is hidden from sight and there are numerous examples of design and education interventions that can bring it to the forefront and intensify the interaction between people and nature. In England, the Canal & River Trust has painted a white line marking duck lanes along waterway paths in London, Birmingham and Manchester to encourage people to be mindful of the waterfowl they share the road with when walking or biking (Werber 2015). In Vigis Park (Vilnius, Lithuania) tiny road signs were created to illustrate areas where wildlife are present, but often remain unseen (Figure 13.3c). Artist/designer Chris Woebken (2012) designed a billboard that serves as a safe roosting habitat for bats (from the *White Nosed Syndrome* threatening American bats) and also picks up bat calls to translate them into text messages on the billboard (Figure 13.3d). Such multifunctional interventions not only expose people to bats, but also demonstrate their ecological function, as important pest control in human habitats. This billboard harnesses the same screen technology that normally distracts people from the experience of nature and demonstrates that with creative design, it can be used to provide knowledge and help reconnect people with nature. Online wildlife guidebooks and applications are increasingly common today. They deliver thousands of images, sounds and information that can help people know about, find and interact with nature in the same way these technologies help people to find their way in cars, choose restaurants and communicate with each other (Beatley 2011).

Notwithstanding traditional conservation education and outreach activities, urbanization and new technologies are providing novel opportunities for experiences of nature. There is no substitute to outdoor activities for establishing meaningful interaction with nature and a biophilic city is an outdoorsy city (Beatley 2011). Outdoor activities such as hiking or gardening that enable children and adults to explore, touch, hear, smell, see and learn about plants and wildlife must be promoted, if we are to ensure experience of nature and its associated benefits. But we must also acknowledge that technology-based interaction allows us to "mediate, augment and simulate the natural world" (Kellert 2014). Although it is not yet clear whether such indirect experiences can introduce profound affection for and concern about nature, nature-based citizen science programs and other online activities are becoming increasingly popular. One example is the diversion of the *Pokemon GO* augmented reality game to promote tangible waste removal in nature (*Pokedechets GO*). Whether such initiatives will generate lasting changes in connection to nature remains an open question. Another example is the Noah's Wild Cities project: Urban Biodiversity and Community that has over 7,000 members worldwide and over 40,000 mapped documentations of wildlife in cities. These are located on an interactive map, also including the possibility to ask the community for help to identify species based on their photos. This should also tap into the idea that exploring and learning about biodiversity can be much more fun at a community or collective level (e.g. BioBlitzes and citizen science; Beatley 2011). In fact, one of the major challenges in enhancing the connection of people with nature is to develop techniques that do not aim to standardize a particular idea or experience of nature, but rather educate people about the different ways to experience it (Clayton *et al.* 2017).

Policy recommendations

Mitigating the extinction of experience cannot be achieved without the support of local and national authorities that can promote various policies to enhance people-biodiversity interaction and find solutions for legal and regulatory constraints. For instance, in sprawled residential areas, tax-incentives can encourage private landowners to implement biodiversity-friendly practices in their properties and maintain connectivity. This can help establish large green areas that can support viable populations of several species. With careful management, motivated landowners can even plant endangered species, or provide nesting habitats for endangered species in their backyard (Marris 2013). In more dense business districts, where open green spaces are scarcer and fragmentation is higher, cities can also use tax-incentives, subsidies, regulatory incentives (e.g. easier permits procedures) or regulations to encourage businesses to increase their green index and connectivity. For example, in 2009, Toronto adopted a legislation that requires the installation of green roofs for all residential and commercial rooftops above 2000 m². Local authorities should also set up strong legislation and enforcement against individuals or businesses that damage habitats and key species. Such actions can reflect the severity of the issue and set a strong social norm. Finally, local authorities should support a network of public and private institutions that will promote outdoors activities and educate children and adults about nature (Beatley 2011). For instance, supporting the growing movement of community gardening or adopting nature reserves in the close vicinity of the city, arranging free activities and transport, can serve as an excellent policy to enhance the people's interaction with nature. It is indubitable that local authorities have it in their hands to kick start and re-create a virtuous cycle of nature experience.

Conclusion

The extinction of experience is a major contemporary social and environmental issue. It leads to negative effects on health and well-being, fundamentally transforms how people value nature and it can undermine conservation efforts (Miller 2005; Dunn *et al.* 2006). Addressing this issue is pivotal for the planning and management of sustainable cities in the future. It calls for guaranteeing that our cities will harbor a complex and biodiverse nature, but also and equally pervasive, that our cities will facilitate access, expose and nudge people to experience, explore, gain knowledge and get connected to nature (Beatley 2011). Such experiences can take a variety of forms and we should also accept and encourage new experiences that are different than the ones of earlier generations (Clayton *et al.* 2017). Designing such biophilic cities can sometimes appear to be conflicting with regional conservation goals, and we need to develop strategic approaches to minimize such instances. Maintaining the love of and affinity to nature of the majority of the world population is vital for stimulating the behavioral change and public support needed to achieve conservation objectives. Therefore, averting the extinction of experience may well be the grandest challenge of conservation scientists and practitioners.

References

Ajzen I. (1991) Theories of cognitive self-regulation: The theory of planned behavior. *Organizational Behavior and Human Decision Processes* 50: 179–211.

Beatley T. (2011) *Biophilic Cities: Integrating Nature into Urban Design and Planning*. Island Press, Washington, DC.

Beninde J., Veith M. and Hochkirch A. (2015) Biodiversity in cities needs space: A meta-analysis of factors determining intra-urban biodiversity variation. *Ecology Letters* 18: 581–592.

Bratman G.N., Daily G.C., Levy B.J. and Gross J.J. (2015) The benefits of nature experience: Improved affect and cognition. *Landscape and Urban Planning* 138: 41–50.

Clayton S., Collenoy A., Conversy P., Maclouf E., Martin L., Torres A.C., Truong M.X. and Prevot A.C. (2017) Transformation of experience: Toward a new relationship with nature. *Conservation Letters* (2017).

Couvet D. and Prevot A.C. (2015) Citizen-science programs: Towards transformative biodiversity governance. *Environmental Development* 13: 39–45.

Cox D.T.C. and Gaston K.J. (2015) Likeability of garden birds: Importance of species knowledge & richness in connecting people to nature. *PLoS ONE* 10, e0141505.

Dallimer M., Irvine K.N., Skinner A.M.J., Davies Z.G., Rouquette J.R., Maltby L.L., Warren P.H., Armsworth P.R. and Gaston K.J. (2012) Biodiversity and the feel-good factor: Understanding associations between self-reported human well-being and species richness. *BioScience* 62: 47–55.

Dunn R.R., Gavin M.C., Sanchez M.C. and Solomon J.N. (2006) The pigeon paradox: Dependence of global conservation on urban nature. *Conservation Biology* 20: 1814–1816.

Fuller R.A. and Gaston K.J. (2009) The scaling of green space coverage in European cities. *Biology Letters* 5: 352–355.

Gladwell V.F., Brown D.K., Barton J.L., Tarvainen M.P., Kuoppa P., Pretty J., Suddaby J.M. and Sandercock G.R.H. (2012) The effects of views of nature on autonomic control. *European Journal of Applied Physiology* 112: 3379–3386.

Goddard M.A., Dougill A.J. and Benton T.G. (2010) Scaling up from gardens: biodiversity conservation in urban environments. *Trends in Ecology and Evolution* 25: 90–98.

Jabareen Y.R. (2006) Sustainable urban forms their typologies, models, and concepts. *Journal of Planning Education and Research* 26: 38–52.

Kellert S.R. (2014) *Birthright: People and Nature in the Modern World.* Yale Press, New Haven, CT.

Lacoeuilhe A., Prévot A.-C. and Shwartz A. (2017) The social value of conservation initiatives in the workplace. *Landscape and Urban Planning* 157: 493–501.

Ledoux C.N. (1804) *L'architecture consideree sous le rapport de l'art, des moeurs et de la legislation.* H. L. Perroneau, Paris.

Lin B.B. and Fuller R.A. (2013) Sharing or sparing? How should we grow the world's cities? *Journal of Applied Ecology* 50: 1161–1168.

Louv R. (2008) *Last Child in the Woods: Saving Our Children from Nature-deficit Disorder.* Algonquin Books, Chapel Hill, NC.

Marris E. (2013) *Rambunctious Garden: Saving Nature in a Post-wild World.* Bloomsbury Publishing, New York.

Metcalfe D. (2015). *Multispecies Design.* University of the Arts London, London.

Miller J.R. (2005). Biodiversity conservation and the extinction of experience. *Trends in Ecology and Evolution* 20: 430–434.

Mora C. and Sale P.F. (2011) Ongoing global biodiversity loss and the need to move beyond protected areas: A review of the technical and practical shortcomings of protected areas on land and sea. *Marine Ecology Progress* 434: 251–266.

Nassauer J.I. (1995) Messy ecosystems, orderly frames. *Landscape Journal* 14: 161–170.

Papworth S.K., Rist J., Coad L., Milner-Gulland E. J. (2009) Evidence for shifting baseline syndrome in conservation. *Conservation Letters* 2: 93–100.

Pett T.J., Shwartz A., Irvine K.N., Dallimer M., Davies Z.G. (2016) Unpacking the people–biodiversity paradox: A conceptual framework. *BioScience* 66 (7): 617.

Pyle R.M. (2003) Nature matrix: Reconnecting people and nature. *Oryx* 37: 206–214.

Pyle R.M. (1978) The extinction of experience. *Horticulture* 64–67.

Shanahan D.F., Bush R., Gaston K.J., Lin B.B., Dean J., Barber E. and Fuller R.A. (2016) Health benefits from nature experiences depend on dose. *Scientific Reports* 6. Available at www.nature.com/articles/srep28551 (accessed 21 August 2017).

Shwartz A., Cosquer A., Jaillon A., Piron A., Julliard R., Raymond R., Simon L. and Prévot-Julliard A.C. (2012) Urban biodiversity, city-dwellers and conservation: How does an outdoor activity day affect the human–nature relationship? *PLoS ONE* 7: e38642.

Shwartz A., Muratet A., Simon L. and Julliard R. (2013) Local and management variables outweigh landscape effects in enhancing the diversity of different taxa in a big metropolis. *Biological Conservation* 157: 285–292.

Shwartz A., Turbé A., Julliard R., Simon L. and Prévot A.-C. (2014a). Outstanding challenges for urban conservation research and action. *Global Environmental Change* 28: 39–49.

Shwartz A., Turbé A., Simon L. and Julliard R. (2014b) Enhancing urban biodiversity and its influence on city-dwellers: An experiment. *Biological Conservation* 171: 82–90.

Soga M. and Gaston K.J. (2016) Extinction of experience: The loss of human–nature interactions. *Frontiers in Ecology and the Environment* 14: 94–101.

Soga M., Yamaura Y., Aikoh T., Shoji Y., Kubo T. and Gaston K.J. (2015) Reducing the extinction of experience: Association between urban form and recreational use of public greenspace. *Landscape and Urban Planning* 143: 69–75.

Soga M., Yamaura Y., Koike S. and Gaston K.J. (2014) Land sharing vs. land sparing: Does the compact city reconcile urban development and biodiversity conservation? *Journal of Applied Ecology* 51: 1378–1386.

Sushinsky J.R., Rhodes J.R., Possingham H.P., Gill T.K. and Fuller R.A. (2013) How should we grow cities to minimize their biodiversity impacts? *Global Change Biology* 19: 401–410.

Ulrich R.S., Simons R.F., Losito B.D., Fiorito E., Miles M.A. and Zelson M. (1991) Stress recovery during exposure to natural and urban environments. *Journal of Environmental Psychology* 11: 201–230.

van den Berg A.E., Maas J., Verheij R.A. and Groenewegen P.P. (2010) Green space as a buffer between stressful life events and health. *Social Science & Medicine* 70: 1203–1210.

Werber C. (2015) London's canal walkways now have "duck lanes". Available at http://qz.com/408647/londons-canal-walkways-now-have-duck-lanes/ (accessed 26 December 2016).

Wilson E.O. (1984) *Biophilia*. Harvard University Press, Cambridge, MA.

Woebken C. (2012). *Bat Billboard*. Available at: http://chriswoebken.com/WORK/BAT-BILLBOARD (accessed 26 December 2016).

14

BIODIVERSITY-FRIENDLY DESIGNS IN CITIES AND TOWNS

Towards a global *biodiversinesque style*

Maria Ignatieva

Introduction

Modern urban landscape design is dominated by conventional, paradigm–creating global homogeneous landscapes using principles of the English style park (often called picturesque), gardenesque (Victorian) and modernist styles. Green areas, such as public parks, private gardens and street plantings, consist of extended manicured lawns with scattered groups of decorative trees and shrubs and flowerbeds. The global style is based on using a pool of 'chosen' plants from different parts of the world, which are grown in global nurseries (Ignatieva 2011). The use of similar design and planning principles for organisation of urban spaces, applying similar plants and construction materials, is leading to loss of local identity and biodiversity and is increasing the cost of maintenance and management (Ignatieva and Ahrné 2013). The reason for creating such homogeneous and unsustainable landscapes is a strong desire to offer a 'familiar', uniformly predictable, prefabricated modernistic vision for green areas in urban spaces that will be acceptable to businesses and urban dwellers. By the beginning of the twenty-first century, certain landscape architecture elements such as lawns had become symbols of the prevalence of market economics and a demonstration of wealth and success. Design and construction companies of all sizes rushed to offer clients this obviously 'successful' green cliché (Figure 14.1).

Interestingly, the eighteenth century English park landscape that the modern world has adopted as its model is based on a picturesque convention of green extended lawns/pastures with tree clumps and lakes that seem intrinsic to nature, but in reality are artificially created. This 'green' landscape has been mistakenly taken to represent ecological quality. The gardenesque garden style added even more artificiality (demonstration of art power over nature) and took urban green areas even further away from real nature. Ironically, modern global green areas based on these two landscape architecture styles, which require great efforts for establishment and maintenance, and are the most powerful and perfect expression

FIGURE 14.1   Landscape of global lawns. Panama City. October 2016. Photo: M. Ignatieva.

of artificiality (Mosser 1999), are nevertheless seen by people as very 'natural' and very 'green' elements.

Questions are being increasingly asked today about how 'green' these green areas really are and about the quality of the green. Existing urban lawns and flower-beds do offer certain ecosystem services, but taking into consideration the essential artificiality and high resource requirement of these urban green areas, their real worth in our cities can be questioned. The search for a new expression of identity of place and the development of a new design language that can allow sustainable urban green areas to be created is currently gaining interest worldwide.

One possible solution in this situation is to develop a new landscape architecture style, *biodiversinesque*, as an alternative to the existing picturesque–gardenesque approach of creating global homogeneous landscapes (Ignatieva and Ahrné 2013). This innovative approach would be based on the new knowledge about ecology and design and involve recent nature–based solutions. The *biodiversinesque style* calls for a new vision of using biodiversity as a means of returning real nature to the city and as a new tool for landscape design.

One of the essential tasks in developing and introducing a new landscape design paradigm is to identify ways of translating ecological patterns and functions related to biodiversity into cultural and aesthetic language that can be understood and accepted by urban citizens.

This chapter discusses several practical examples from Europe (Great Britain, Sweden and Germany), New Zealand and China, where biodiversity is used as an important landscape design tool.

Design with native plants

A native component of biodiversity (native flora and fauna) is seen as an especially powerful way to express urban ecological and cultural identity in New Zealand. It is a country that has experienced tremendous pressure from the introduction of exotic plants and animals during the past 150 years. The Englishness of urban landscapes, dramatic loss of natural ecosystems and introduction and naturalisation of thousands of exotic plants (today the number of completely naturalised exotic species in New Zealand exceeds the number of indigenous plants) has made protection and restoration of biodiversity a high priority. New Zealand today even has to use the terms 'native biodiversity' or 'indigenous biodiversity'. Native biodiversity is distinguished from the exotic (non-indigenous, non-native) component of New Zealand biodiversity that is based on plants living outside their native distributional range and that has arrived in New Zealand as a result of human activity, either deliberately or accidentally. The speed at which the native New Zealand biota has been suppressed and changed is unprecedented (Meurk 2007). In the past 30 years there has been a battle to shift the emphasis of landscape architects and planners to designing with native plants. New Zealand's vision to shift its paradigm to using native species is also important for indigenous Maori people in preserving their ancestor's lands and traditions. Loss of native plants and plant communities, and dominance of agricultural and urban landscapes with a high level of exotic plant invasion make such an indigenous biodiversity approach very welcome.

Design with native plants in New Zealand emerged through a series of overlapping stages. Robinson (1993) suggested extracting *plant signatures* from particular native plant communities and adapting them to existing (even very modified) conditions, thus offering unique experiences of local nature. Lucas Associates (1997) published a series of maps for Christchurch based on analysis of environmental conditions prior to the development of colonial settlements and offered private gardeners, planners and designers a choice of key indigenous plants and associated bird species. In the past decade, there have been a growing number of publications on the themes of *Living with Natives* (Frey and Spellerberg 2008) and *Design with Natives* (Spellerberg and Frey 2011) that show the results of practical work by garden designers and landscape architects in New Zealand and discuss new ways of thinking and understanding the new paradigm of designing with New Zealand's native plants. There has also been experimental research resulting in suggestions on using these native plants for lawns, hedges, shrubs and tree groups in private and public gardens (Ignatieva *et al.* 2008). Today native New Zealand plants are offered in local nurseries and shopping centres throughout New Zealand. An increasing number of nurseries specialise entirely in growing species from nearby habitats, thus offering a local genetic pool of plants.

The main concept in the latest design movement is to understand the unique nature of New Zealand and to find ways of incorporating it into an urban setting, in order to strengthen the sense of place and local character. Interestingly, however, New Zealand designers are using quite a limited number of native species and rely mostly on the philosophy 'keep it simple, Nature will do the rest'. Thus, they use repetition of biodiversity features and the simplicity principle, which provides rest for the viewer's eyes (Blakely 2011). The majority of landscape architects who have decided to go along the path of designing with nature give priority to indigenous plants with distinctive form, texture and colour, which makes their designs clearly naturalistic and different from conventional modernistic landscape design (Figure 14.2). However, there is also a group of designers who believe in the use of a mixture of native and non-native trees, shrubs and some perennials that can provide colour and seasonal interests and reflect another layer of New Zealand's history – the colonial era. These designers aim to respect and reflect the complicated New Zealand urban landscape narrative.

FIGURE 14.2 Design with native plants in one Christchurch neighbourhood. Photo: Maria Ignatieva.

One of the most striking examples of employing the theme has been the use of New Zealand plants in a very symbolic demonstration of nature and urban community resurrection in Christchurch. In restoration work after the devastation caused by the earthquake on 22 February 2011, native plants have been used (Figure 14.3). Pocket gardens are filling the gaps left by the removal of ruined buildings all over the city centre. There is also a plan for the Avon–Otakaro network which aims to use released areas (after the removal of destroyed houses) for creating a nature reserve. Here the goal is 'to turn a tragedy into an opportunity, a polluted drain into a vibrant river system, and exhaustion and despair into hope and inspiration' (www.avonotakaronetwork.co.nz, accessed on 27.12.2016).

Biodiversity as a main design tool: alternative lawns

Lawns are one of the most common elements in modern cities, covering up to 70 per cent of open urban space (in public and private gardens, cemeteries, roadsides, golf courses and sports fields). This commonality of modern lawns makes them an 'essential' element in urban landscapes, without any questioning of their meaning and values. Since lawns are green, this automatically makes them very 'natural' and desirable. However, the artificiality, use of industrial homogeneous monoculture mixtures and intensive management practices used on lawns, such as frequent mowing and spraying of herbicides and fertilisers, have raised concerns about their potential negative impact on the urban environment (Ignatieva *et al.* 2015).

FIGURE 14.3 Memorabilia with native plants: remembering a destroyed Christchurch Cathedral. Photo: Maria Ignatieva.

English approach

England is considered one of the cradles of conventional lawns and the 'giver' of the picturesque and gardenesque styles, which are based on smooth green carpets. However, even British gardeners have started to question this approach and to promote enrichment of existing grasslands (including lawns) with various native and exotic species. Woudstra and Hitchmough (2000) point out that Britain has a long and rich history of enriching lawns with native and exotic herbaceous plants and provide some examples from medieval (flowery mead) and even baroque gardening. The most influential figure in this regard was William Robinson with his ideas of 'wild gardens' in the late nineteenth and early twentieth century. He tried to emulate natural groupings (including wild flowers) and promoted the use of hardy bulbs (native and exotic) in lawns or meadows, with the intention of giving a more natural, 'wild' appearance to plantings. He also called for wiser use of lawns and a reduction in contemporary mowing practices, in order to let flowering plants blossom and be enjoyed by the public (Woudstra and Hitchmough 2000).

Naturalistic plantings

The late twentieth and early twenty-first century saw the development of the Sheffield School of planting design within British landscape architecture. This was based on the work of James Hitchmough and Nigel Dunnett and their rich horticultural British experience and contemporary ecological and botanical research, particularly on the approach of dynamic plant communities (influenced by life strategies (Grime 1974) and disturbance). The main aim of Hitchmough and Dunnett was to develop a new generation of urban plantings that 'exploit ecological as well as horticultural processes and understanding' (Hitchmough and Dunnett 2004, p. 2). This vision calls for low maintenance costs and biodiversity (which supports maximum wildlife), as well as aesthetically acceptable plantings. The main distinctive feature of the Sheffield approach is respect for 'traditional horticultural wisdom' and trying to find a compromise between using native and exotic herbaceous species. This is very understandable considering the long horticultural tradition of using and trying new exotic species that has been the essence of British gardening, with its unique climate and economic conditions (British Empire with its great colonial opportunities).

The new British philosophy of 'nature-like planting' or 'naturalistic plantings' has been identified as adopting one of three approaches (Hitchmough and Dunnett 2004). *Habitat restoration* is based on existing, local species, using local, extant populations that occurred on the site in the past. It is heavily based on using only native plants. However, some point out the difficulties of applying this 'purist' approach to urban landscapes. *Creative conservation* uses the concept of planting native species that can be obtained from native plant nurseries within the geographical region. Finally, the *anthropogenic landscape approach* is based around species 'that could never have "naturally" occurred on the site but which may, given its current conditions, be well fitted to it' (Hitchmough and Dunnett 2004, p. 7). This approach is based

on plant communities that include native and also exotic species. Such communities can be even completely novel and never have existed before in any region.
While applying the same ecological processes and principles as habitat restoration and creative conservation, the anthropogenic landscape approach very much
allows and even encourages the use of non–native species. The main justifications
and reasons for wider implementation are plant relevance for a particular site, natural selection and the opportunity for low maintenance. Another crucial criterion in
the anthropogenic landscape approach is consideration of the aesthetic perspectives
and acceptance by the public and managers.

British promoters of anthropogenic naturalistic vegetation are very much driven
by issues of biodiversity and sustainability in general. They also desire to revisit and
re-evaluate existing semi–natural and spontaneous urban plant communities and see
them as environmentally friendly and inexpensive replacements for traditional turf
(Dunnett 2008). They also seek to use natural and semi–natural plant communities
(especially meadows, steppe, prairie and woodland edges), as well as spontaneous
vegetation, as an inspiration for further design manipulation guided by ecological and
aesthetic principles and aiming to address the sense of local place. Strong lobbying
on the 'rightness' of using non–native plants in naturalistic plantings is justified by the
'very small native flora' (Hitchmough and Dunnett 2004, p. 10). That is why there is
a perceived need to add colours, textures and forms that are absent in native English
plant communities, but are supported by centuries of successful British horticulture.

Anthropogenic naturalistic plantings include quite a range of types. The most
common are 'naturalistic herbaceous plantings', which include different ranges of
vegetation. For example, they include some native meadows, which are mostly
grass-dominated communities with some common forbs. However, the main
focus of the Sheffield school is on the development of different mixed native-
exotic meadows (sown matrices of native grasses and different herbaceous species
with added planted exotic forbs from the Himalayas/East Asia, Caucasus or US)
(Kingsbury 2004; Hitchmough 2004; 2009).

Annual plantings (pictorial meadows) made from seed mixtures of native and
exotic colourful annuals and some biennials or short-lived perennials (*Papaver
rhoeas, Agrostemma githago, Anthemis arvensis, Centaurea cyanus, Linum grandiflorum,
Eschsholzia californica, Coreopsis tinctoria,* etc.) are one of the most common and popular types, due to their visual impact, ease of establishment and relatively low cost.
Lately, the biodiversity factor (attractiveness to pollinators) is also gaining increasing weight in the argumentation on using annual plantings in urban environments.

The culmination of the Sheffield naturalistic plantings school of Hitchmough
and Dunnett came in the planting concept for London's Queen Elizabeth
Olympic Park. Numerous pictorial meadows and naturalistic herbaceous plantings
were realised on a tremendous scale of 25 hectares. There were yellow and gold
'Olympic Gold Meadows', European meadows and planting inspired by the temperate Americas, temperate Asia, the Mediterranean and Asia Minor and even the
Southern Hemisphere (particularly South Africa) and different types of wildflower
meadows (Figure 14.4).

FIGURE 14.4 Naturalistic plantings in London's Olympic Park. Photo: Maria Ignatieva.

Particularly after the success of the Olympic Park planting, Sheffield's 'naturalistic plantings' have become one of the main streams of ecological design in Britain and Europe, and very influential and even fashionable in other parts of the world.

Grass-free lawns

British horticulturist Lionel Smith (Smith and Fellowes 2014) came up with the idea of 'grass-free lawn', which is made up of specific mowing-tolerant, low-growing plants instead of grass. In other words, the proposed perennial grass-free communities are mown only a few times a season, expelling the principle of the 'snobbish' English green 'carpet'. Smith believes that grass-free lawns can be a good substitute for traditional lawns because they are environmentally friendly (less energy input in maintenance and a biodiversity-rich plant community), while also providing 'cues to care' (the intention of human presence and care, which should be visible) (Nassauer 1995). Grass-free lawns are inspired by the medieval idea of flower-rich turf and meadows in Great Britain, which, as mentioned earlier, has a long history of enriching grassy turf surfaces using exotic perennials. Smith suggests removing the monoculture of grasses and replacing it with a polyculture of perennial species that have decorative foliage and flowers (native and exotic). Growing such grass-free/tapestry lawns in England would require over 50 per cent less mowing than conventionally managed lawns (Smith and Fellowes 2014). Tapestry lawns can thus be a good option for areas in private or public parks that would otherwise be managed as monotonous green lawn.

Apart from a desire to use biodiversity as an important design tool and an ambition to decrease the costs of maintenance, the major driving force behind grass-free lawns is to create an aesthetic appearance for the ground surface that is accepted by the general public as well as managers, since it offers both attractive flowers and diverse foliage forms. This approach is thus similar to 'naturalistic plantings' but is very much British in its essence, since it is based on the experience of successful horticulture and the British passion for gardening.

Alternative lawns in Sweden

Swedish lawns were recently researched within the transdisciplinary project 'Lawns as an ecological and cultural phenomenon. Searching for sustainable lawns in Sweden' (Ignatieva et al. 2015). Sweden has a long history of garden development. The country has always had close connections with different European countries and has adopted influential garden styles. Cut turf from surrounding meadows embedded with native grasses and forbs were probably used in medieval castles and monasteries. Decorative grass surfaces were used in Renaissance and Baroque formal, as well as in landscape (English)-style Swedish parks. The nineteenth century public parks signified the era of triumphal use of lawns as a major element of urban open spaces. Introduction of the Swedish model (which was a result of the activities of the Social Democratic Party from 1932 to 1976 and a desire to create an equal society) resulted in wide use of modernistic planning and architectural values. Thus, lawns were a standard element that fitted perfectly into the functionalistic aesthetics of a simplified, rationalistic (prefabricated) style (Ignatieva et al. 2017).

Some researchers (Woudstra 2004) decoded the influence of William Robinson and Gertrude Jekyll in some garden plantings in post-World War II Sweden (naturalised tulips in long grass, streams and meadow plants). Holger Blom (head of Stockholm's park administration from 1938 to 1971) was a very important figure in the implementation of green areas policies and practices in Sweden. He established regular connections with different countries, particularly Great Britain. It is not surprising that naturalistic plantings (particularly pictorial meadows; Figure 14.5) were so well received in Swedish municipalities (Tregay and Gustavsson 1983; Woudstra 2004).

Despite the dominance of lawns in Swedish public parks and multifamily residential areas (built during the People's Homes and Million Homes programmes) during the twentieth century, there were several early examples of implementation of authentic Nordic ideas in the development of lawns. For example, some native landscapes with semi-natural vegetation (forests, woodlands, meadows and pastureland) were preserved in newly created parks and other green areas (Florgård 2009). This was part of the Nordic School and Nordic classicism (Nolin 2004).

The use of native meadows instead of lawns was a very important concept in the Stockholm School of Parks (developed in the 1930s and 1940s). This school advocated a new park style, in contrast to the dominant conventional modernistic approach (Florgård 1988). Native species and wildflower meadows were used in

FIGURE 14.5 Pictorial meadow in Örebro municipality, June 2015. Photo: I. Runeson.

addition to conventional lawns. One of the classical examples of a park of this kind is Norr Mälarstrand in Stockholm.

The great expense of conventional lawn maintenance and the associated negative environmental impact has forced a search for alternative solutions to lawns. However, changing the existing paradigm of 'ideal' lawn strategy and implementation of new approaches requires special planning and design solutions adjusted for each particular case.

In our project in Sweden, we studied lawns from different perspectives as a social (cultural perceptions of lawns) and ecological phenomenon (biodiversity, vascular plants, pollinators and earthworms), in order to understand their roles in sustainable urban planning, design and management. One of the goals of the project was to devise a new generation of alternative lawns for Sweden.

Our vision of alternative lawns is to create biodiverse, aesthetically pleasing and cost-effective plant communities based on the diverse Swedish flora. The inspiration for this approach came from the Swedish firm *Pratensis*, a pioneer in conservation of natural Swedish grasslands and promoting the use of biodiverse alternative solutions for lawns. The firm uses only seeds collected from natural sources (native plant communities within different parts of Sweden). *Pratensis* is thus a genetic 'purist' and the Swedish alternative approach to conventional lawns is grounded on local plants that are extremely cost-efficient and sustainable in the northern climate. In addition, this approach corresponds to the character of meagre Nordic nature with its modest colour and texture. In contrast, the British promote a more 'pictorial' approach of using exotic colourful herbaceous plant material (for example from American prairie, South Africa or Asian grasslands) because of people's 'love' of colour.

Inspired by existing examples of *Pratensis* activities in different Swedish cities, we created the Experimental Alternative Lawn Trial in Ultuna Kunskapspark (Knowledge Park at SLU, Uppsala Campus) as an important educational facility for academics and public communities. Here, visitors can see alternative lawn solutions (Figure 14.6). We created different plots: grass-free lawn, a bumblebee paradise, chalky meadow, gravel lawn, grass-free meadow, dry meadow and a butterfly paradise. The bumblebee paradise solution was created using a mixture of species that are attractive to bumblebees, for example *Echium vulgare* and *Galium verum*. A more extreme example is the gravel lawn, which was established on the toughest soil with a high gravel content using plants such as *Antennaria dioica*, *Hypochaeris radicata* and *Trifolium arvense* that are proving very successful on this soil.

FIGURE 14.6 Demonstration of alternative lawns (inspired by Demonstration Trial in Ultuna Campus) in Sundbyberg, Stockholm. Photo: Maria Ignatieva.

Inspired by the British concept of grass-free/tapestry, we also introduced two Swedish versions of this type of alternative lawn. Our Swedish grass-free (tapestry) lawn references and is inspired by the medieval idea of the *Garden of Eden* (paradise), with an informal 'flowery mead' or meadows planted with a great variety of aromatic herbs and flowers. In medieval times, such flowery meadows were about enhancement of natural beauty in the direction of religious meaningfulness, where each plant has symbolic meaning. In modern times, our task is similar to that with old gardens, namely using flowering plants to create a *locus amoenus* or delightful place, a new urban place where man and wild nature (plants, insects and birds) can coexist (Naydler 2011; Buizer *et al.* 2016). The task of our Swedish grass-free lawn is to underline modern ecological 'symbolism' by improving biodiversity and returning to nature. This lawn consists of 29 herbaceous plants native to Sweden which can provide the effect of a low-growing flowering carpet that can be used for recreation and which will be cut only 2–3 times during the summer season (Figure 14.7).

FIGURE 14.7   Grass-free lawn on Ultuna Campus, June 2016. Photo: Maria Ignatieva.

In the social part of the project on Swedish lawns, we also surveyed urban dwellers in multifamily housing areas in regard to their perceptions of alternative lawns. People were asked to comment on three types of alternative lawns, illustrated by pictures (tapestry/grass-free lawns with low-growing herbaceous plants, meadows with perennials framed by conventional short-cut lawns, or meadows with annuals (pictorial meadow)). The respondents expressed quite a range of opinions. One group would like to see flower-rich meadows and understood the environmental value and cost-effectiveness of such meadows. However, many people still preferred more conventional lawns but also argued that meadows could be established in some designated places in a neighbourhood. The most interesting finding was appreciation of grass-free lawns, which in some places people called 'amazingly beautiful'. The only concerns about such grass-free lawns were a fear of damaging the flowering plants. Perennial meadows framed by mown grass areas received positive feedback from respondents in many neighbourhoods. This confirms the effectiveness of the 'cues to care' approach (Nassauer 1995), when people can easily see use of design and care and can accept 'messy' aesthetics. People even expressed a desire to have such framed meadows in unused places (Ignatieva *et al.* 2017).

Analysis of historical precedents and modern trends in landscape architecture indicates the importance of searching for a new paradigm of alternative lawns in Sweden. While considering the importance of the 'cues of care' approach and acknowledging people's attachment to green manicured carpets, it is important to find ways of appreciating nature as it is, without reinforcing its colours or forms. Creation of sustainable and cost-effective urban landscapes cannot be achieved without understanding and appreciation of common, even very 'meagre' or 'messy' native and spontaneous ecosystems.

Changing aesthetics: appreciation of wastelands and 'go spontaneous' in Germany

The German approach to engaging urban biodiversity in design is well known among urban ecologists and landscape architects. Because of the wide-scale destruction during the Second World War, Germany has experience of managing an abundance of derelict sites and wastelands. Spontaneous vegetation is a characteristic component of its urban ecosystems. It is no coincidence that West Berlin became the cradle of urban ecology. Surrounded by the Wall, the city was forced to deal with abandoned railways, ruins and other urban habitats occupied by spontaneously growing vegetation. The famous example of Natur-Park Südgelände (abandoned railroad yard) in Berlin illustrates where nature and naturally occurring processes were accepted, valued and used as a fundamental landscape design tool. In these areas, spontaneous vegetation was primarily used as a symbol of the past and thus connected to its authenticity (Kuhn 2006). At the same time, such sites were seen as places for keeping and maintaining rich urban biodiversity resources and a species gene pool and, in a context of rapid urbanisation and dramatic loss of natural ecosystems, as an important place for experiencing nature in the city. Appreciation and understanding of the important role of spontaneously occurring urban habitats even resulted in integration of such sites into urban planning documents and their acceptance as specific sites/examples of urban post-industrial novel ecosystems.

In the past two decades, the spontaneous vegetation approach has moved up to a new level and now German landscape architects and urban ecologists are designing with spontaneous vegetation. For example, in Berlin's multifunctional park Gleisdreieck (opened in 2011), which was established on the wasteland of a former railway yard, some areas were purposely left to 'go wild' and to allow spontaneous vegetation to colonise gravel and rubble (Figure 14.8). Other areas of the park are covered by different kinds of sown meadows. These 'wild', nature-like pieces are deliberately contrasted with conventionally designed lawns and recreational facilities.

One of the more important features of the spontaneous design paradigm is the belief that 'If it is possible to make spontaneous vegetation more attractive, it may also be possible to introduce it as an alternative to ornamental planting in the city' (Kuhn 2006, p. 46). This clearly indicates that the design and aesthetic

FIGURE 14.8 Spontaneous vegetation colonisation in action. *Solidago canadensis* (exotic plant that escapes from cultivation and now one of the most common species of wastelands in Germany) is one of the very successful and competitive species in rubble debris in Gleisdreieck Park, Germany. Photo: Maria Ignatieva, 2016.

component is coming to the fore and becoming the main argument for promoting these spontaneously appearing urban plant communities, as an appropriate alternative to gardenesque ornamental city plantings. This approach is based on the idea of 'improving' existing spontaneous plant communities by adding more flowering species (native or exotic). The latest research on making spontaneous vegetation more visible to urban citizens involves using steppe and prairie species (Köppler et al. 2014), which is in line with the British anthropogenic naturalistic planting.

A relaxed and flexible approach of using native and non-native plant species in spontaneous vegetation is based on ecological conditions: spontaneous plants are already there and well adapted to given urban conditions. There are a small number of threatened naturalised exotic species in German urban flora. Another reason behind such an approach is related to German history. Prioritising the use of native plants in landscape design was 'high-jacked' by National Socialism, which rejected the exotic and prescribed use of indigenous plants in order to achieve 'natural design', which was seen as the most significant characteristic of German landscape design (Wolschke-Bulmahn 1997).

Due to rapid loss of natural vegetation, especially grasslands, there is a growing body of research on propagating native plants and producing meadow seed mixtures that can be used in urban environments in Germany. For example, demonstration meadows have been established at the University of Applied Sciences in Erfurt and there are a growing number of newly established sustainable complexes (e.g. UmweltBundesamt Building (The Federal Environment Agency (UBA), Dessau) that actively promote the concept of urban biodiversity, including the use of native species in the design of different urban habitats and mixtures from local brownfield sites.

Kongjian Yu's 'big feet' ecological design

One of the first landscape architects in China to introduce a pioneering alternative vision to existing global homogenisation was Kongjian Yu, with his new paradigm of 'big feet'. The ancient tradition of foot binding in China, which sacrifices the function of rustic 'big feet' in the name of gentrification and beauty, was chosen as the main motto of this new Chinese landscape architecture approach. Its principles are opposed to contemporary Chinese global urban design landscapes, which are viewed as too costly (financially and environmentally) and unsustainable. Yu is one of the harshest critics of the Chinese 'City Beautiful' central government gardening approach (Yu and Padua 2007), stating that 'Torn between its own imperial past and today's Westernization, what is China's identity?' (Yu 2006, p. 9). Most of contemporary Chinese landscape architecture practice is heavily reliant on the baroque, picturesque and gardenesque styles that are often mixed with traditional Chinese aesthetics (for example rockeries copied from Chinese classical gardens). Yu also opposed the expansive and intensively managed exotic decorative species that are completely devoid of native vegetation. When China entered the global market economy it began to produce a prefabricated Chinese version of global landscapes. These landscapes have completely lost an important spiritual connection to ancestors' places and Chinese identity (Yu 2006).

Yu's metaphor of celebrating the aesthetic of high-performing, low-cost, healthy feet resulted in the creation of an *adaptive landscape design*, which is based on traditional Chinese farming technology combined with principles of ecological design. The aim of each project is to provide an 'environment with self-sustaining identity' (Yu 2012, p. 72).

Key terms in the Yu landscape design philosophy are 'bigfoot aesthetics', 'beauty of weeds', 'new vernacular' and 'ecological minimalism' (Yu 2006; Sounders 2012). Native plants play a crucial role in Yu's projects, especially at wetland and river restoration sites (Figure 14.9).

His projects favour untrimmed 'weedy' native, low-maintenance plants, especially in post-industrial landscapes, as well as plants that can adapt to water level fluctuations (for wetland and waterway restoration projects). One very important principle of Yu's alternative design is to make land 'productive', which is why traditional crop species are actively used in his projects (e.g. Turenscape, Shenyang Architectural University Rice Campus, Shenyang City, Liaoning Province, China, Shanghai Houtan Park) (Figure 14.10).

FIGURE 14.9 Shanghai Houtan Park. Using native plants for water purification and river/pond restoration. Photo: Maria Ignatieva, April 2015.

FIGURE 14.10 Shanghai Houtan Park. Adaptive design by Kongjian Yu. *Leucanthemum vulgare* is used as a kind of crop (designed meadow). This plant has a peak blossom in the end of April–May. Photo: Maria Ignatieva, April 2015.

The Yu approach makes ecological processes quite visible by using apposite landscape design language and reinforcing or contrasting natural and man-made elements (for example the red ribbon design, which became Yu's 'design signature'). Protection of biodiversity is seen as an important objective, to be provided together with other ecologically driven solutions (stormwater management, flood control, water treatment). Even though it prioritises the use of native species, the Yu approach cannot be seen as 'purist', since if a design solution requires the use of certain exotic plants to reinforce the aesthetic perception, these are allowed in certain places, in monoculture or in mixtures with a more natural appearance.

Discussion

In this chapter, we discussed the consequences of creating similar global urban landscapes and the use of biodiversity as an important and powerful tool of returning nature to cities. We also highlighted the need for developing a new *biodiversinesque style* that allows the translation of urban ecological patterns and functions into a design language that can be accepted and understood by urban dwellers.

Great Britain followed its tradition of being a global garden trendsetter by being the first to suggest a new ecological approach of 'naturalistic plantings' that is based on knowledge of ecological patterns and long horticultural traditions of working with exotic plants. This approach has been accepted and used in some European countries, for example Sweden. On the other hand, New Zealand, which in past centuries created 'mirror' motherland gardenesque-picturesque landscapes, at the end of the twentieth century chose its own way – using native biodiversity as a major design tool in returning lost native landscapes and native identity.

The New Zealand pathway of preferring and promoting native biodiversity is well justified, since that country has experienced dramatic environmental crises, loss of native ecosystems and a tremendous amount of invasive exotic species. For New Zealand, design with native plants is probably the only way of saving unique flora and genus loci.

The German approach of using particularly spontaneous vegetation for decorative purposes is novel and respectful of the history of a site. This vision is economically profitable, respects existing ecological conditions and, since it demonstrates nature and ongoing natural processes (such as succession), can connect urban people to nature. Even though it is very much German in its nature and closely related to German history and planting design, the latest trend for spontaneous vegetation design and the idea of improving its 'weedy' appearance by planting beautiful flowering prairie or steppe species is very similar to the British anthropogenic naturalistic planting, where landscape architecture employs horticulture principles (colour, texture, form) to please the eye of Western European city dwellers. Otherwise, it is believed that wild, messy and weedy nature would not be 'welcome' in an urban environment.

The real value of the alternative ecological design by Kongjian Yu in China is his call for complete opposition to existing Westernisation and his proposal of a complex approach based on interaction of natural, cultural and spiritual processes that respect Chinese traditions. In this case there is a distinct emphasis on the value of vernacular landscapes. It seeks to go back to the value and cultural identity of the landscape and recover past 'survival skills' for dealing with flooding, soil erosion and drought. This approach is quite naturally based on the use of indigenous plant material. However, Yu's vision is 'design with nature'. Thus, even when the design is based on ecological processes and respect for local environmental conditions, there are still modern design skills to make this landscape look interesting and acceptable to modern urban dwellers. The designer in such an approach is a nature 'improver', who can decide, for example, that a sunflower crop, perennial meadow or monoculture of *Lythrum salicaria* will look extremely beautiful to visitors during the flowering period. The next step in developing such innovative, nature-based solutions may be to identify additional models from lost natural landscapes and bring them to urban environments.

The problem facing landscape architecture worldwide today is the danger of unintentionally transferring new ecological design solutions to other countries without critical evaluation of their appropriateness and suitability for local conditions. Of course the exchange of ideas in landscape architecture is unavoidable and there is a place where 'naturalistic plantings', spontaneous vegetation or grass-free lawn can be established, for example in Chinese or Japanese public gardens. However, there is an urgent need for research on different approaches that can be suitable for each country. The meaning of 'naturalness' and 'human influence' varies in different cultures (Jorgensen 2004). However, new ecology-based solutions developed in Western countries must not be seen as absolute and introduced in practice without consideration of local history and ecological conditions. In global urban landscapes, such as in Western Europe and the US, people are more connected to other places than to their neighbours and their local environment.

A different approach (plurality) in searching for a local *biodiversinesque style* should be an important part of the research in ecological planting design. The realm of global 'flow' should be shifted to the local and the regional. Another important direction in developing a *biodiversinesque approach* is education and cooperation with social media, which can help urban practitioners and dwellers to accept and appreciate real green, i.e. nature 'in the raw', rather than 'improved' or 'manicured' green.

References

Blakely P. (2011) Our philosophy is keep it simple: Nature will do the rest. In Spellerberg I. and Frey M. (eds), *Native by Design: Landscape Design with New Zealand Plants*, pp. 8–19, Canterbury University Press, New Zealand.

Buizer M., Elans B. and Vierikko K. (2016) Governing cities reflexively: The biocultural diversity concept as an alternative to ecosystem services. *Environmental Science and Policy* 62: 7–13.

Dunnett N. (2008) *Pictorial Meadows*. In Müller N., Knight D. and Werner P. (eds), Urban Biodiversity & Design: Implementing the convention on biological diversity in towns and cities. Third conference of the Completence Network Urban Ecology, Erfurt, 21–24 May 2008. Book of Abstracts. BfN, Bonn, 64.

Florgård C. (1988) Det långsamma skådespelet. *Utblick Landskap* 2: 36–39.

Florgård, C. (2009) Preservation of original natural vegetation in urban areas: An overview. In McDonnell M.J., Hahs A.K. and Breuste J.H. (eds), *Ecology of Cities and Towns: A Comparative Approach*, pp. 380–398. Melbourne: Australian Centre for Urban Ecology.

Frey M. and Spellerberg I. (2008) *Living with Natives: New Zealanders Talk About Their Love of Native Plants*. Canterbury University Press, Christchurch, NZ.

Grime J.P. (1974) Vegetation classification by reference to strategies. *Nature* 250: 26–31.

Hitchmough J. (2004) *Naturalistic Herbaceous Vegetation for Urban Landscapes*. In (Dunnett N. and Hitchmough J. (eds), *The Dynamic Landscapes*, pp. 130–183. Taylor & Francis, London.

Hitchmough J. (2009) Diversification of grassland in urban greenspace with planted, nursery-grown forbs. *Journal of Landscape Architecture* 4(1): 16–27.

Hitchmough J. and Dunnett N. (2004) Introduction to naturalistic planting in urban landscapes. In Dunnett N. and Hitchmough J. (eds), *The Dynamic Landscapes*, pp. 130–183. Taylor & Francis, London.

Ignatieva M. (2011) Plant material for urban landscapes in the era of globalisation: Roots, challenges and innovative solutions. In Richter M. and Weiland U. (eds), *Applied Urban Ecology: A Global Framework*, pp. 139–161. Blackwell, Oxford.

Ignatieva M., Meurk C., van Roon M., Simcock R. and Stewart G. (2008) *Urban Greening Manual. How to Put Nature in our Neighbourhoods: Application of Low Impact Urban Design and Development (LIUDD) principles, with a biodiversity focus, for New Zealand developers and homeowners*. Landcare Research Sciences series no.35. Manaaki Whenua Press, Lincoln, N.Z.

Ignatieva M. and Ahrné K. (2013) Biodiverse green infrastructure for the 21st century: From 'green desert' of lawns to urban biophilic cities. *Journal of Architecture and Urbanism* 37(1): 1–9.

Ignatieva M., Ahrné K., Wissman J., Eriksson T., Tidåker P., Hedblom M., Kätterer T., Marstorp H., Berg P., Ericsson T. and Bengtsson J. (2015) Lawn as a cultural and ecological phenomenon: A conceptual framework for transdisciplinary research. *Urban Forestry & Urban Greening* 14: 383–387.

Ignatieva M., Eriksson F., Eriksson T. Berg P. and Hedblom M. (2017) The lawn as a social and cultural phenomenon in Sweden. *Urban Forestry & Urban Greening* 12: 213–223.

Jorgensen A. (2004) The social and cultural context of ecological plantings. In Dunnett N. and Hitchmough J. (eds), *The Dynamic Landscapes*, pp. 293–322. Taylor & Francis, London.

Kingsbury N. (2004) Contemporary overview of naturalistic planting design. In Dunnett N. and Hitchmough J. (eds), *The Dynamic Landscapes*, pp 58–96, Taylor & Francis, London.

Köppler M.R.R., Kowarik I., Kühn N. and von der Lippe M. (2014) Enhancing wasteland vegetation by adding ornamentals: Opportunities and constraints for establishing steppe and prairie species on urban demolition sites. *Landscape and Urban Planning* 126: 1–9.

Kuhn N. (2006) Spontaneous vegetation as the basis for innovative green planning in urban areas. *Journal of Landscape Architecture* 1(1): 46–53.

Lucas Associates (1997) *Native Ecosystems of Otautahi Christchurch. Set 1–4. For Christchurch-Otautahi Agenda 21 Committee*. Lucas Associates, Christchurch, NZ.

Meurk C. (2007) Implication of New Zealand's unique biogeography for conservation and urban design. In Stewart M., Ignatieva M., Bowring J., Egoz S. and Melnichuk I. (eds), *Globalisation of Landscape Architecture: Issues for Education and Practice*, pp. 142–145. St Petersburg's State Polytechnic University Publishing House, St Petersburg, Russia.

Mosser M. (1999) The saga of grass: From the heavenly carpet to fallow fields. In Tessyot G. (ed.), *The American Lawn*, pp. 40–63, Princeton Architectural Press, Princeton, NJ.

Nassauer J.I. (1995) Messy ecosystems, orderly frames. *Landscape Journal* 14(2): 161–170.

Naydler J. (2011) *Gardening as a Sacred Art*. Floris Books, Edinburgh.

Nolin C. (2004) Public parks in Gothenburg and Jönköping: Secluded idylls for Swedish townfolks. *Garden History* 32(2): 197–212.

Robinson N. (1993) Place and plant design: Plant signatures. *The Landscape*, Autumn, 53: 26–28.

Smith L. and Fellowes M. (2014) The grass-free lawn: Management and species choice for optimum ground cover and plant diversity. *Urban Forestry and Urban Greening* 13(3): 433–442.

Sounders W. (2012) Ecology, with pleasure. In Saunders W. (ed.) *Designed Ecologies: The Landscape Architecture of Kongjian Yu*, pp. 8–9. Birkhauser, Basel, Switzerland.

Spellerberg I. and Frey M. (2011) *Native by Design. Landscape Design with New Zealand Plants*. Canterbury University Press, Christchurch, NZ.

Tregay R. and Gustavsson R. (1983) *Oakwoods New Landscape: Designing for Nature in the Residential Environment*. Alnarp and Runcorn: Swedish University of Agricultural Sciences rapport S&L 15, and Warrington and Runcorn Development Corporation.

Wolschke-Bulmahn J. (1997) Nationalization of nature and naturalization of the German nation: "Teutonic" trends in early twentieth-century landscape design. In Wolschke-Bulmahn J. (ed.), *Nature and Ideology: Natural Garden Design in the Twentieth Century*. pp. 187–291, Dumbarton Oaks, Washington, DC.

Woudstra J. (2004) The changing nature of ecology: A history of ecological planting (1800–1980). In N. Dunnett and J. Hitchmough (eds), *The Dynamic Landscapes*, pp. 58–96, Taylor & Francis, London.

Woudstra J. and Hitchmough J. (2000) The enmelled mead: History and practice of exotic perennials grown in grassy swards. *Landscape Research* 25(1): 29–47.

Yu K. (2006) *Position Landscape Architecture: The Art of Survival*. China Architecture and Building Press, Beijing, China.

Yu, K. (2012). The big feet aesthetic and the art of survival. *Architectural Design* 82(6): 72–77.

Yu K. and Padua M.G. (2007) China's cosmetic cities: Urban fever and superficiality. *Landscape Research* 32(2): 255–272.

15

INTEGRATING URBAN BIODIVERSITY MAPPING, CITIZEN SCIENCE AND TECHNOLOGY

Cynnamon Dobbs, Angela Hernández, Francisco de la Barrera, Marcelo D. Miranda and Sonia Reyes Paecke

Urban biodiversity mapping

In the urban context the most commonly mapped feature related to natural resources is green spaces. This feature is recognized as the existing habitat for several species in urban areas and it is valued for its potential to foster biodiversity (Ignatieva *et al.* 2011; Beninde *et al.* 2015). Urban green spaces can foster a variety of species, from common and abundant species to rare species, and can range from remnant forests and grasslands to vacant lots, parks, community gardens and cemeteries (Jarvis and Young 2005). But not only green spaces can be mapped, other features therein such as vegetation, waterways and abundance of flora and fauna can also be mapped. Mapping in urban areas integrates advanced technologies and methods with those typically used in the field, and needs to combine the skills of a multidisciplinary team including GIS and remote sensing experts, ecologists and social scientists (Pilughe *et al.* 2016).

In order to create maps for urban biodiversity it is necessary to have adequate data that need to be frequently updated. The most commonly used data consists of: 1) data derived from remote sensors, aerial photographs and radar such as vegetation indices and 2) *in situ* data, for example flora and fauna censuses, urban forest inventories and animal counts.

Remote sensing data

This type of data can represent coarser scale indicators of biodiversity information, mainly related to land cover/use, habitat, green spaces or vegetation that can be linked to species presence or diversity. The advantage of this data is that it can be obtained and replicated at different spatio-temporal scales and extrapolated to a longer area than the actual data. The disadvantage is that it requires knowledge on the processing of remote sensing data and species presence cannot be obtained directly from it, even with very-high resolution data. The type of data obtained from these sources typically corresponds to land use/cover maps and derived products from remote sensing imagery such as vegetation indices and topography maps.

Images at high and medium spatial resolution are the most commonly used for urban research and applications. High resolution satellite-borne sensors (<10 m) include IKONOS, QuickBird, GEOEYE, WorldView, OrbView-3 and SPOT-5. The advantage of these images for urban biodiversity mapping relies in their precision in the identification and characterization of small objects, which can lead to the identification of small green spaces and their structure which is essential for habitat suitability measures. Mid-resolution images (10–100 m), such as Landsat, SPOT HRV and ASTER, are able to detect larger green spaces but have the advantage of providing a temporal continuity of information, allowing exploratory and temporal studies to be developed, besides being freely available. Large spatial resolution images, such as those obtained from MODIS, are less frequently used. However, these are useful to assess temporal changes of the whole urban area despite being less adequate for urban biodiversity studies.

Some of the most commonly used resources derived from remote sensors imagery are vegetation indices, particularly for biodiversity purposes. A full review of vegetation indices can be found in Jensen (2007). Of all of those, the most frequently used index is the Normalized Difference Vegetation Index (NDVI). This index combines the use of the red and near infrared bands from airborne imagery to detect the vegetation portion of a landscape, which usually ranges between values of 0.3 to 0.8 (Stathopoulou et al. 2007). This index allows the detection of green spaces and can be used as input for the classification of vegetation.

Land use/cover data can be used for predicting vegetation structure, productivity and landscape metrics enabling the quantification of biodiversity changes or the effects of disturbances (e.g. habitat loss and fragmentation) (Hedblom and Mörtberg 2011). However, the resolution related to land use/cover is not as useful in the urban environment where disturbances, differently from natural environments, occur at much finer scales. Mid-resolution data, such as that derived from Landsat, is one of the most commonly used sources for obtaining landscape structure, habitat quality or vegetation types at a spatial scale good enough to detect variations in the urban area (Hedblom and Mörtberg 2011).

Remote sensing data can also be used as input for habitat suitability models based on environmental data (Martinez-Harms et al. 2016), which can be assessed for predicting single species habitat suitability at a regional scale or in areas that are subjected to urbanization (Hedblom and Mörtberg 2011). In addition, structural connectivity can be determined using cost–distance models or graph theory (e.g. GUIDOS software, Conifor Sensinode).

Summarizing, data from remote sensors by itself can be used for a variety of analysis that can contribute to the knowledge on urban biodiversity such as:

1) Spatial distribution of green spaces can be digitized from satellite imagery or aerial photographs, or can be derived from the calculation of vegetation indices such as the NDVI (Kong and Nakagoshi 2006).

2) Vegetation associated to green spaces, streets, residential areas, among others can be derived in high resolution from vegetation indices or laser-scansions (LiDAR) data in combination with land use cover (Alonzo *et al.* 2014).

3) Monitoring of the spatio-temporal dynamics of remnant areas by using mid-resolution imagery and obtaining the NDVI can enable the detection of changes in the city and assess the effects of urbanization on remnant forests.

4) Classification of different types of vegetation through remote sensing can be used to differentiate between type of vegetation e.g. forests, grasslands, woodlands (Hedblom and Söderström 2008).

5) Recognition of green corridors for ecological connectivity though satellite imagery (Hepcan 2013).

In situ data

With these types of data we can represent the composition and spatial distribution of species and therefore estimate the urban biodiversity of an area. These types of data have the advantage of having finer resolution, thus allowing the inclusion of species difficult to monitor with remote sensing techniques (e.g. herbaceous plants, fauna), get their abundance data and calculate biodiversity indices. The biggest disadvantages of these data are the costs for the data collection (i.e. money, time) and the need for specialized or trained people for data collection. Also, data obtained from different sources might not be comparable, given differences in the methods used to collect them. The resulting maps are presence–absence maps or diversity index maps that can be illustrated in variable size grids according to the density of the species count (Figures 15.1a, 15.1b). These maps can be laid over other environmental variables to detect what urban areas support biodiversity hot spots.

In situ data can be used together with remote sensing data for mapping urban biodiversity. For example, data from different species counts was obtained from publicly available databases from Melbourne, Australia, in order to explore species distribution within the City of Melbourne, an administrative unit located in the center of Melbourne. Data from the forest inventory was used to show the distribution of tree species in the municipality in order to detect areas of tree monocultures and areas that are more diverse and therefore more resilient (Figure 15.1a). The abundance of bird species in a 50 m grid using counts of birds collected in a 20–year time window (Figure 15.1b) was then superimposed on that of trees. Structural habitat connectivity can be assessed by using tree crowns polygons obtained from remote sensing or aerial photographs as input for *Guidos*, a freely available fragmentation software (http://forest.jrc.ec.europa.eu/download/software/guidos/). This software enables areas and corridors to be detected that are of importance to maintain connectivity in the landscape (Figure 15.2a). Diversity indices can be quantified in a certain grid size, as shown for the Shannon diversity index of tree species based on the tree inventory (Figure 15.2b).

(a)

(b)

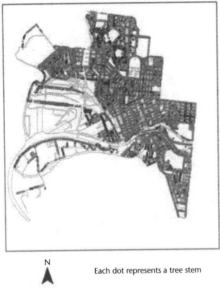

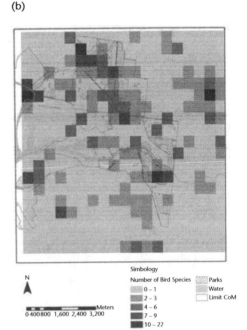

N

Each dot represents a tree stem

Meters
0 500 1,000 2,000 3,000

N

Meters
0 400 800 1,600 2,400 3,200

Simbology
Number of Bird Species   Parks
0 – 1   Water
2 – 3   Limit CoM
4 – 6
7 – 9
10 – 22

FIGURE 15.1 Examples of biodiversity mapping in City of Melbourne, Melbourne, Australia.
a) Urban tree species, b) Bird species richness aggregated by pixel (50x50m
resolution). Courtesy of City of Melbourne.

Use of urban biodiversity maps

As previously mentioned, maps can be a good tool for communicating information
and raising awareness on urban biodiversity in participatory planning programs.
Maps can be integrated into policy and planning documents, either for summarizing main information or for illustrating future interventions, improvements and
zoning. Examples of this can be found in urban biodiversity strategies and planning documents such as the *Urban Biodiversity Strategy of City of Harmony*, Australia
which includes maps of biodiversity corridors, biogeographical zones and biologically significant sites (City of Boroondara 2012). Other examples are found in
State of the Environment reports, such as the vegetation maps included in the
Townsville State of the Environment (www.soe-townsville.org/biodiversity/),
or urban canopy maps such as the USDA Forest Service reports (www.national
geographic.com/news-features/urban-tree-canopy/).

Citizen science for urban biodiversity

As stated above, a significant disadvantage of *in situ* data collection is its cost. This
can at least in part be remedied by involving local residents in the work. *Citizen*

(a)   (b)

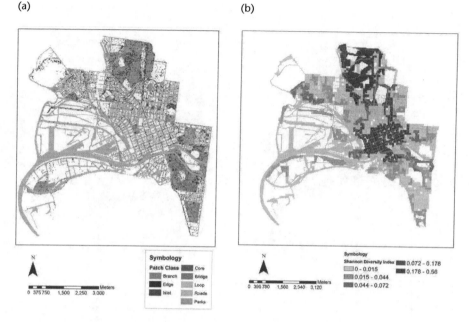

FIGURE 15.2 a) Structural connectivity for urban trees and b) Diversity index for the urban forest. Courtesy of City of Melbourne.

science, referred to as volunteers that participate in scientific studies as field assistants or analysts (Hochachka *et al.* 2012), has been demonstrated as a powerful tool for data-intensive capture across broad spatial and temporal extents phenomena. These types of programs mainly focus on data collection that can offer new opportunities to researchers through web-based tools. Many research projects related to plant and bird biodiversity, tree diseases, infrastructure monitoring, meteorology, social media and mapping have benefited from this kind of program. *In situ* data for bio-diversity mapping can be also gathered using citizen science programs.

One of the main advantages of citizen science relates to the capacity of cap-turing and analyzing large amounts of data over vast geographic areas and long time periods, surpassing the human and technical installed capacity, both in budget and time, available from local and regional governments (Hochachka *et al.* 2012). Other advantages refer to the social benefits related to community empowerment, the democratization of information, social bottom–up integration, social cohesion, knowledge dissemination and community self-education.

Citizen science is an old activity born in the middle of the nineteenth cen-tury when the first data collection was done by European naturalists around the world (Silvertown 2009). Its global expansion occurred with the development of the Information Technology (IT), and especially the Word Wide Web, database systems, smartphones and web mapping. Nowadays, millions of volunteers world-wide work together with scientist and technicians generating new databases that allow new perspectives of analysis, as well as unexpected and innovative results in a variety of areas.

Several steps have been recognized to design and implement a citizen science program (Bonney *et al.* 2014), including:

1) Choosing a scientific or social question, considering that most participants will be amateurs. Here researchers, managers and staff should work together to ensure the social relevance and clarity of the question to be posed and ensure the correct communication of objectives and scope to volunteers and other authorities involved in the project.

2) Building a team consisting of scientists, educators, technologists and evaluators. This multidisciplinary team ensures that protocols of data capture and analysis are adequate for participants and that address the natural variability in the knowledge generated. Usually a citizen science project in relation to urban biodiversity provides an opportunity for developing a multidisciplinary team that includes ecologists, geographers, sociologists and information technology staff.

3) Recruiting participants. The effort in recruitment will depend on the target audience, complexity, spatial and time coverage of the project. To retain volunteers interested in the project some sort of reward is usually needed, such as highlighting their contribution to the advancement the scientific knowledge, visualizing their data on maps and graphs, and showcasing their results on interactive platforms. Another strategy is to establish alliances with civil organizations as project partners, such as community organizations, non-profit environmental agencies, after school programs, among others.

4) Training participants. The success of a citizen science program depends on how participants capture data in the field. It is necessary to provide all the support that the participants require to understand the objectives and methodologies associated with the project. The use of IT based platforms such as Skype, Facebook, Bloggers, Wordpress among others, allows building widely spread programs for training and evaluation. E-learning platforms are educational capital for this process.

5) Accepting, editing and displaying data. All the information generated must be accepted, edited and made available for analysis, not only by professional scientists but also by the general public. Both robust protocols for data control and adequate IT infrastructures are necessary to ensure a robust process of data capturing.

6) Analyzing and interpreting data. Due to citizen science producing coarse data sets, analyses of the data should be oriented towards the recognition of emerging spatial and temporal patterns. Methodologies coming from data mining frameworks are good alternatives for this purpose. A variety of commercial and open source software is today available for analysis, such as RAPIDMINER (https://rapidminer.com/), SPSS (www.ibm.com/analytics/us/en/technology/spss/), WEKA (www.cs.waikato.ac.nz/ml/weka/) and R Data Mining (www.rdatamining.com/).

7) Disseminating results. Different alternatives for communication must be considered, including scientific journals, technical reports for target audiences, on line newsletters and blogs, project websites among others. Experience has shown that newsletters (printed or online) are effective tools to communicate research progress, to report the recruitment of new volunteers and lay audiences, and also to enable participants to communicate their own results or reflections (Hochachka *et al.* 2012; Newman *et al.* 2012). This is a fundamental step in citizen science programs; recognizing the contribution of volunteers in the program as well as serving as an education tool.

8) Measuring outcomes. Measuring the impact of a citizen science project is one of the most important challenges that include assessing the effects in different social dimensions. The level of engagement of volunteers, the interaction and feedback between volunteers and researchers, the reaching of results to the urban population and decision makers are some of the dimensions that need to be considered.

One of the complexities of these programs is the uncertainty of the data; this has been overcome by new methodologies including complex model design, protocols for data acquisition and control of process (from the field work to the results dissemination) and a high level of training for volunteers. A common approach is data verification by experts in order to detect possible errors in the records (Lucky *et al.* 2014). Another approach consists of the preparation of detailed protocols for data acquisition and control.

Some examples of successful citizen science programs include the monitoring of bumblebee (*Bombus sp.*) populations in UK cities, ant species diversity and abundance in US cities, and urban bird diversity in the USA, the UK and Canada (McCaffrey 2005; Lye *et al.* 2012; Lucky *et al.* 2014). Also in recent years, the development of numerous applications for smartphones has allowed the desired data to be easily entered into customized forms and the location and time of the citizen science data to be automatically registered (Dickinson *et al.* 2012; Newman *et al.* 2012). Such is the case of the *eBird project* (www.eBird.org), where the data must be entered into forms containing a defined list of species that can be spotted at a given location, decreasing the chances of misidentifications and errors. If a user wants to report a species that has not been registered in a certain region or season, he must follow a series of additional steps to enter the data and allow its validation from a researcher (Hochachka *et al.* 2012). Other examples include the www.treezilla.org app for capturing tree data and their benefits which are delivered in a map platform; www.opentreemap.org, a collaborative platform for crowdsourcing tree inventories, ecosystem service quantification, urban forestry analysis and community engagement; and *ENV,* a new smartphone application for recording wildlife crimes. Citizen science projects provide a good opportunity to collect data from private land, by recruiting landowners to conduct urban wildlife monitoring (e.g. arthropods, insects, birds) in their own backyards and gardens (Cooper *et al.* 2007).

Another successful project is the *Tucson Bird Count*, a volunteer based program for urban birds monitoring in Tucson, Arizona. This project was first focused on measuring the abundance and distribution of native birds in various habitats across the city, but its long-term goal is to identify the necessary conditions to sustain native birds (McCaffrey 2005). The project has generated a detailed database and distribution maps of 212 species of birds, with more than 164,000 records in a 5-year project. Their results have been used in the local land-use planning and to evaluate potential sites for new urban parks (McCaffrey 2005).

Citizen science projects can also contribute to monitoring invasive species. In Chile, the *Harlequin Ladybeetle Project* has enabled the monitoring of a rapidly spreading invasive species, *Harmonia axyridis*, throughout the country (www.chin ita-arlequin.uchile.cl). The project combines monitoring with education about the ecological importance of native lady beetles, and the dissemination of the invasive control measures. The project also maintains a website in which citizens can submit their data, indicating the approximate number of individuals observed, their activities and habitat. The project contains an interactive map for registering the location (Grez and Zaviezo 2015).

As was exposed in this section citizen science projects can be delivered in a mapping platform and can use technology for gathering and disseminating results. The following section details the use of technology for gathering urban biodiversity data.

Integrating technology for urban biodiversity

Technology can aid in monitoring and communicating urban biodiversity and can be a useful tool to collect citizen science data. Monitoring urban biodiversity requires continuous updating that can be gathered using technology such as apps and online platforms, the same forms that can be used for communicating information to the general public. Technology can also be incorporated for information gathering of urban biodiversity data from the citizen scientists either from an app platform, online mapping platform or image collection.

Apps for tablets and mobile devices

The appearance and availability of smartphones and tablets have greatly facilitated the design of numerous apps that serve for biodiversity purposes. First, they serve to track urban biodiversity (monitoring) and feed collaborative databases by taking advantage of the geo-location capacities incorporated in mobile devices. They serve for connecting people to urban biodiversity by facilitating data collection from citizens. In both cases the data can be delivered real-time after some steps of validation and processing of the collected data. Most of the built apps are for Android and Apple users, and a few of them can be used without mobile devices (e.g. *iTree* and *eBird*). The apps also can be used for education purposes, looking at creating awareness of children and other citizens first, and second, for science purposes. A good example is the *Urban iNaturewatch Challenge* conducted in India with the aim of "building field-oriented environmentally resilient communities" (www.inaturewatch.org). The apps linked to urban biodiversity can be classified into four groups:

1) Urban forests, trees and other vascular plants represent the type of species most commonly seen and tracked. In addition, many citizens hold "very strong personal ties to urban trees and forests" (Dwyer *et al.* 1991, p. 15). This has led to the appearance of a wide variety of apps facilitating the identification of trees and, to a lesser extent, other plants (e.g. *iKnow Trees 2 LITE, Trees, Tree Identification, Tree Id – British Trees, Trees of Britain, PlantNet, Smart'Flore*). These apps include photos for each species and detailed descriptions of the plant structures, and their locations. Some of the apps are limited to certain regions, usually because they are developed from local government initiatives (e.g. *TreeMapLA* for greater Los Angeles, *Urban Forest Map* for San Francisco, *iTrees* for four Indian cities). Finally, a few of them provide information on the environmental benefits of urban forests (e.g. *Urban Forest Map, iTree*).

2) Birds and bird-watching is an activity that claims the attention of a lot of people. Birds are highly appreciated and known by people not involved in science (Chace and Walsh 2006). Several apps contain data to help in their identification, including sounds and songs (e.g. *Merlin*). Some apps even allow automatic recognition of birds by recording their songs (e.g. *Warblr*). Many of them are geographically restricted to certain areas, for example *Bird Id – British Birds* for the UK, *All Birds Germany!* for Germany and *iNaturewatch birds* for India. Most of the technologies associated to bird watching, identification and tracking are not specially designed for urban birds but are still utilized in urban environments. The *eBird* platform (ebird.org) is the main initiative for birds tracking. Since its launch in 2002, it has become the biggest collaborative database of bird abundance and distribution over time, covering mostly natural but also urban environments. There are also events that connect apps to citizen science such as *Celebrate Urban Birds* (celebrateurbanbirds.org) and the *Great Backyard Bird Count* (gbbc.birdcount.org). *Celebrate Urban Birds* is a permanent program focused on 16 species of birds common in the US, Canada and Mexico, involving people from children to the elderly to participate in scientific investigations by collecting high-quality data about these species. The *Great Backyard Bird Count* takes place on only one weekend a year. In 2016, it involved more than 130 countries counting more than 5,500 species of birds.

3) Insects are also a group quite easily collected and recorded in urban and natural environments through the use of these technologies. Some examples of mobile apps are *iRecord Butterflies* (UK), *iRecord Ladybirds* (UK), *iRecord Grasshoppers* (UK), *iButterflies* (India).

4) Invasive species is a group especially important in cities because urban ecosystems include several exotic species that might become invasive. Science

and practitioners have seen the opportunity of involving citizens for gathering timely information about the presence of invasive species and have asked for help by using apps such as *Early Detection Network series (US)* which is linked to some programs focusing on urban areas (e.g. metroinvasive.info), or *Outsmart Invasive Species* (US) designed for tracking invasive plants and insects throughout Massachusetts. Other examples are *AquaInvaders* (Scotland) and *Whatsinvasive*, with many data tracked in urban areas, and the *iPest* series designed for dealing with US urban pests.

Map online platforms

The information collected by citizen scientists can feed map online platforms. Citizen scientists can deliver their data in a traditional way (e.g. full worksheets), automatically from apps or by using new online platforms that allow entering new data. More importantly, citizens can access a large volume of data and get informed and educated by themselves, thus not only providing information. They can visualize raw and scientifically processed data and even edit data provided by other users. One example of this is the *OpenTreeMap* (opentreemap.org). This is an initiative that facilitates the collaborative creation of urban forest maps, the performance of some analyses and the exploration of the benefits provided by them (ecosystem services). It has been applied mostly in the US and UK, mainly in urban areas. The above mentioned *eBird* platform allows people's own data to be submitted and also to explore automatically created maps reporting biodiversity hotspots (ebird.org/ebird/hotspots). Also, the platform allows maps to be built based on queries from specific species (ebird.org/ebird/map). They have data distributed worldwide and some data within cities.

It is not only species that can be collaboratively mapped—also, potential habitats and patches of urban ecosystems are mapped by citizen scientists. An example is the *Habitat Network* (content.yardmap.org) that aims to map backyards, parks, farms, favorite birding locations, schools and gardens.

Data to build maps on urban biodiversity can be derived from two main sources, *in situ* data and technological applications. Technology data can come from remote sensing or photointerpretation sources, while *in situ* data can be collected by experts or citizens and can be uploaded onto technological platforms or gathered on paper (Figure 15.3). The use of remote sensing is becoming more frequent in urban studies, given its rapid data capture, interpretation and temporal availability of data. Similarly, data obtained from citizen science is also becoming popular, given the ability to create public awareness, educate and generate community engagement.

Biodiversity mapping can inform planners and educators, create public awareness, assist in infrastructure building, and illustrate information on species presence, abundance, diversity indices and landscape fragmentation. This will improve current and future decision making towards building sustainable cities and conserving urban biodiversity and ecological processes.

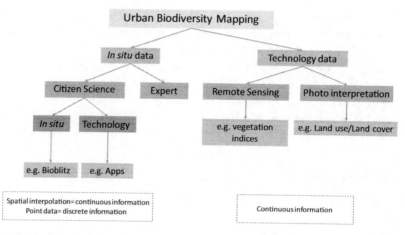

FIGURE 15.3 Process for urban biodiversity mapping using *in situ* data and technology derived data.

References

Alonzo M., Bookhagen B. and Roberts D.A. (2014) Urban tree species mapping using hyperspectral and lidar data fusion. *Remote Sensing of the Environment* 148: 70–83.

Beninde J., Veith M. and Hochkirch A. (2015) Biodiversity in cities needs space: A meta-analysis of factors determining intra-urban biodiversity variation. *Ecology Letters* 18: 581–592.

Bonney R., Shirk J.L., Phillips T.B., Wiggins A., Ballard H.L., Miller-Rushing A.J. and Parrish J.K. (2014) Next steps for citizen science. *Science* 343: 1436–1437.

Chace J.F. and Walsh J.J. (2006) Urban effects on native avifauna: A review. *Landscape and Urban Planning* 74(1): 46–69.

City of Boroondara (2012) Urban Biodiversity strategy 2013–2023. Borrondara, City of Harmony. Available at www.boroondara.vic.gov.au/sites/default/files/2017-05/Urban-Biodiversity-Strategy-2013-2023.pdf (accessed 21 August 2017).

Cooper C.B., Dickinson J., Phillips T. and Bonney R. (2007) Citizen science as a tool for conservation in residential ecosystems. *Ecology and Society* 12(2): 11.

Dickinson J.L., Shirk J., Bonter D., Bonney R., Crain R.L., Martin J., Phillips T. and Purcell K. (2012) The current state of citizen science as a tool for ecological research and public engagement. *Frontiers in Ecology and the Environment* 10(6): 291–297.

Dwyer J.F., Schroeder H.W. and Gobster P.H. (1991) The significance of urban trees and forests: Toward a deeper understanding of values. *Journal of Arboriculture* 17: 276–284.

Grez A.A. and Zaviezo T. (2015) Chinita arlequín: *Harmonia axyridis en Chile*. Available at: www.chinita-arlequin.uchile.cl (accessed 29 October 2016).

Hedblom M. and Söderström B. (2008) Woodlands across Swedish urban gradients: Status, structure and management implications. *Landscape and Urban Planning* 84: 62–73.

Hedblom M. and Mörtberg U. (2011) *Characterizing Biodiversity in Urban Areas Using Remote Sensing*, in Yang X. (Ed) Urban remote sensing: Monitoring, synthesis and modeling in the urban environment. First Edition, John Wiley & Sons Ltd.

Hepcan S. (2013) Analyzing the pattern and connectivity of urban green spaces: a case study of Izmir, Turkey. *Urban Ecosystems* 16: 279–293.

Hochachka W.M, Fink D., Hutchinson R.A., Sheldon D., Wong W.K. and Kelling S. (2012) Data-intensive science applied to broad-scale citizen science. *Trends in Ecology & Evolution* 27(2): 130–137.

Ignatieva M., Stewart G.H. and Meurk C. (2011) Planning and design of ecological networks in urban areas. *Landscape and Ecological Engineering* 7: 17–25.

Jarvis P.J. and Young C.H. (2005) The mapping of urban habitat and its evaluation. University Wolverhampton, Wolverhampton. Available at www.ukmaburbanforum.co.uk/documents/papers/MABpaper.pdf (accessed 21 August 2017).

Jensen J.R. (2007) *Remote Sensing of the Environment*. 2nd edn. Pearson, Prentice Hall, Upper Saddle River, NJ.

Kong F. and Nakagoshi N. (2006) Spatial–temporal gradient analysis of urban green spaces in Jinan, China. *Landscape and Urban Planning* 78(3): 147–164.

Lucky A., Savage A.M., Nichols L.M., Castracani C., Shell L., Grasso D.A., Mori A. and Dunn R.R. (2014) Ecologists, educators, and writers collaborate with the public to assess backyard diversity in The School of Ants Project. *Ecosphere* 5(7): 1–23.

Lye G.C., Osborne J.L., Park K.J. and Goulson D. (2012) Using citizen science to monitor Bombus populations in the UK: Nesting ecology and relative abundance in the urban environment. *Journal of Insect Conservation* 16(5): 697–707.

Martinez-Harms M.J., Quijas S., Merenlender A.M. and Balvanera P. (2016) Enhancing ecosystem services maps combining field and environmental data. *Ecosystem Services* 22: 32–40.

McCaffrey R.E. (2005) Using citizen science in urban birds studies. *URBANhabitats* 3(1): 70–86. Available at: www.urbanhabitats.org/v03n01/citizenscience_full.html (accessed on 29 October 2016).

Newman G., Wiggins A., Crall A., Graham E., Newman S. and Crowston K. (2012) The future of citizen science: Emerging technologies and shifting paradigms. *Frontiers in Ecology and the Environment* 10(6): 298–304.

Pilughe G., Fava F. and Lupia F. (2016) Insights and opportunities from mapping ecosystem services of urban green spaces and potential in planning. *Ecosystem Services* 22: 1–10.

Silvertown J. (2009) A new dawn for citizen science. *Trends in Ecology & Evolution* 24: 467–471.

Stathopoulou M., Cartalis C. and Petrakis M. (2007) Integrating Corine land cover data and Landsat TM fir surface emissivity definition: Application to the urban area of Athens, Greece. *International Journal of Remote Sensing* 28(15): 3291–3304.

INDEX